ALSO BY MARK JONES LORENZO

Affront to Meritocracy: Stories of Overlooked Talents,
Ignored Abilities, and Hidden Truths

Not Ok: A Requiem for GW-BASIC

Apophenia's Antidote: A Probability and Statistics Primer

Its Wildness Lies in Wait: Mathematical Fallacies, Cognitive Traps,
and Debunking the Myths of the Lottery

the paper computer unfolded

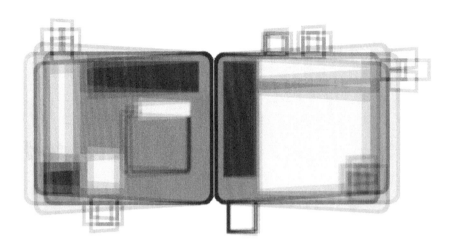

the paper computer unfolded

A TWENTY-FIRST CENTURY GUIDE TO

the bell labs CARDIAC
cardboard illustrative aid to computation

the LMC
little man computer

and the IPC
instructo paper computer

WRITTEN AND ILLUSTRATED BY
MARK JONES LORENZO

SE BOOKS
Philadelphia | Pittsburgh

* The entities associated with the instructional models described herein did not review or endorse this book.

Ψ
SE BOOKS
5307 West Tyson Street
Philadelphia, Pennsylvania 19107
www.sebooks.com

Copyright © 2017 by Mark Jones Lorenzo

All rights reserved. Printed in the United States of America. No part of this book may be reproduced in any manner whatsoever without written permission except in the case of brief quotations embodied in critical articles and reviews. For information, contact SE BOOKS.

References to websites (URLs) were accurate at the time of writing. Neither the author nor SE BOOKS is responsible for URLs that may have expired or changed since the manuscript was prepared.

Published in full-throated defiance of Yog's Law.

Cover design and art, as well as all illustrations in the text, by Mark Jones Lorenzo.

Microsoft Excel is a registered trademark of the Microsoft Corporation.

The respective entities, inventors, creators, current or former copyright or trademark holders, or proprietors associated with the instructional models described hereafter did not review, authorize, or endorse any of the contents of this book.

Library cataloging information is as follows:

Lorenzo, Mark Jones
 The paper computer unfolded : a twenty-first century guide to the Bell Labs CARDIAC (cardboard illustrative aid to computation), the LMC (little man computer), and the IPC (instructo paper computer) / Mark Jones Lorenzo.
 p. ; cm.
  Includes bibliographical references.
  I. Title
 1. Computers— history. 2. Computer programming. 3. Coding theory.
  QA76.17 C22 2017
  004.0920'22053— js22   20176322256
 ISBN: 978-1-537-42113-1

10 9 8 7 6 5 4 3 2 1

These are fast-moving times, and those who make no effort to understand computers may very well get left behind.

— David W. Hagelbarger, writing in 1968

In the world of the future most people will be computer literate.

— Fred C. Matt, writing in 1979

brief contents

introduction ♦ 3

PART A. computers and their instructional paradigms

SECTION 1. the basics of computers ♦ 37

SECTION 2. the cardiac and the little man computer ♦ 77

SECTION 3. the instructo paper computer ♦ 119

SECTION 4. instructional models and the future ♦ 145

PART B. pulp and digital emulation

SECTION 5. the pink paper computer ♦ 157

SECTION 6. a cardiac-little man emulator ♦ 171

SECTION 7. a cardiac-little man assembler ♦ 195

SECTION 8. an instructo emulator and complier ♦ 207

SECTION 9. a turing machines emulator ♦ 237

epilogue: completely unfolded ♦ 243

selected bibliography ♦ 245

about the author ♦ 247

introduction

Bell Telephone Laboratories was the birthplace of many technologies that power our world: the binary digital computer; the transistor, which ultimately led to the microchip; the laser; the solar battery cell; radio astronomy; the C programming language and UNIX operating system; fiber-optic transatlantic cable; and information theory. For a long stretch of the twentieth century, Bells Labs was the birthplace of the future.

 Yet Bell Labs' birth was incidental. After Alexander Graham Bell founded what would become the American Telephone and Telegraph Company (AT&T) in 1870, subsidiaries of the large corporation were established. At around the same time, a company producing electric equipment called the Western Electric Company was incorporated in Chicago. After being sold to Western Union, Western Electric became a pawn in a pitched legal battle between Western Union and AT&T; as a result of a settlement, Western Electric became a subsidiary of AT&T.

 By the early 1900s, Western Electric established a research laboratory, which quickly grew to thousands of people by the mid-twenties— at which point it was spun off, becoming Bell Labs, jointly owned by AT&T and Western Electric, and centralized in New York City. Later, as its growth continued, facilities for research were established in Murray Hill, New Jersey, as well.

 Just as important as what the scientists and the mathematicians at Bell Labs invented, discovered, and improved was their promotional efforts describing the nature of their work. Bell Labs had an intimate relationship with the United States government, through rigging early communication technologies— in 1927, for example, famously transmitting long-distance images of then-Secretary of Commerce Herbert Hoover— to later working hand-in-hand with the military to help secure Allied victory in World War II. And through it all, the public at large came to know and admire Bell Labs' reputation; by the middle of the twentieth century, an idée fixe took hold:

that Bell Labs was reaching ahead of the curve, inventing a beneficent, World's Fair-type future for us.

The public imagination of the amazing innovations at Bell Labs— this notion of Bell Labs as an idea factory, as author Jon Gertner calls it in his definitive institutional history of the same name (wisely calling on the factory-model metaphor that most Americans were familiar with courtesy of the Model T)— could be fed and sustained by big, public, important projects, but also by targeting the young. As part of that effort, Bell Labs distributed in effect self-promotional advertisements in the form of free scientific kits to schools and students nationwide. A flyer for these kits (sometimes also referred to as "units") set the stage. The header featured a large-font quote from the executive secretary of the National Science Teachers' Association: "Put together by scientists and science teachers, these resources offer educators a unique opportunity to expand their effectiveness." Wares containing devices to demonstrate wave motion and double as audio-oscillators are hawked on the flyer, along with repeated pleas to immediately contact "your local telephone company"— Bell Labs was, for decades, an quasi-independent subsidiary of the monopolistic AT&T, albeit with tremendous resources and dominance over the industrial-research landscape— while the "limited quantity" of supplies lasted.

Another contemporaneous advertisement made Bell Labs' intentions even clearer. Leading with the headline "How the Bell System is helping develop gifted scientists and engineers," the sales pitch borders on agitprop:

> Tomorrow's top scientists and engineers are hidden in high school classrooms today. The problem is to find them, inspire them. And Bell System is helping this national effort with a unique series of teaching aids....
>
> The program will continue, with the cooperation of leading educators, as long as it serves a useful purpose. And the Bell System will benefit only as the nation benefits— from better teachers and able young engineers and scientists.

Scientifically literate representatives of Ma Bell's local companies would trek to nearby high schools to demonstrate how to work the kits. An issue of *The Journal of the Telephone Industry* dated February 27, 1965, notes that the "'Aids to High School Science' program is intended as a concrete contribution to education in areas where Bell scientists are particularly competent." These high school "aids" focused on scientific concepts such as solar energy (the Bell Solar Energy Experiment/From Sun to Sound), speech synthesis (An Experiment in Electronic Speech Production), and crystals and light (Experiments with Crystals and Light), among several others. The *Journal* reports that over 120,000 of the kits had been distributed to high school teachers by early 1965. (Bell Labs didn't restrict their attention to high schools: teachers and students in the elementary grades were subject to their benevolent indoctrination, too.)

Many of the Bell Labs' kits came packaged with short films, such as the 15-minute "The Genesis of the Transistor,"[*] a documentary complementing their semiconductor kit, which "tells the story of the discovery of the 'transistor effect' and subsequent invention of the transistor." The film details how semiconductors like germanium split the difference between good conductors like copper, gold, and silver, and poor conductors (and good insulators) like rubber, and increase their conductivity when heated or exposed to light. (The best conductors consist of a handful of elements in which the orbiting electrons are easily dislodged; the better the conductor, the more easily electrons "flow" through the material.) After the war, scientists at Bell Labs turned their attention toward exploiting the properties of semiconductors. In the short film, Walter Brattain, with chalk in hand at a blackboard, explains how he and John Bardeen arrived at a successful semiconductor amplifier; Brattain also describes the day in 1947 when the transistor effect was observed for the first time. Years later, William Shockley improved the transistor. All three scientists later shared the Nobel for their efforts.

The transistor, which supplanted the ungainly and inefficient vacuum tube, is arguably the greatest invention to ever come out of Bell Labs. It's also the most controversial, involving jealousy and betrayals that are at least as interesting as the sold-state physics behind the breakthrough. Shockley had hired Brattain and Bardeen, among other top minds, right after the Second World War to investigate semiconductors. In 1945 Shockley thought he had figured out how to design a semiconductor amplifier, using what he called a "field effect." But it didn't work. A frustrated Shockley assigned Brattain and Bardeen to investigate. And that's where the problems began.

Brattain and Bardeen quickly became a dynamic duo, complementing each other's strengths: one built the experiments, the other interpreted the results. A couple of years passed, and Brattain and Bardeen, sans Shockley, designed a point-contact transistor—which scrapped Shockley's concept in favor of a different approach.

When Shockley found out, he was angry. He felt he had been cast aside, with the spotlight now instead shining brightly on his subordinates. In retaliation, Shockley secretly added to Brattain and Bardeen's design, sketching out and constructing an improved device called the junction (or sandwich)

---

[*] https://www.youtube.com/watch?v=LRJZtuqCoMw

transistor. Shockley had found a way to take credit for a working transistor after all.

As a result of his antisocial behavior Shockley quickly became persona non grata at Bell Labs, so he quit and formed a company. Shockley Semiconductor was the first of the major high-tech companies to populate Silicon Valley, located in the San Francisco Bay Area. In fact, Shockley's belief that silicon was a better choice of semiconductor than germanium helped to ensure that a "Germanium Valley" never appeared on a California map.

Eventually Shockley's brash and rude management style caught up with him. A handful of his top employees, who later became known as the "traitorous eight," left Shockley Semiconductor to start a new company: Fairchild Semiconductor. And from Fairchild, Bob Noyce and Gordon Moore, who were among the traitorous eight, founded Intel Corporation. Of course, Moore is best known for Moore's Law, his prescient prediction about semiconductors: "The complexity for minimum component costs has increased at a rate of roughly a factor of two per year. Certainly over the short term this rate can be expected to continue, if not to increase." And Noyce, along with Jack Kilby at Texas Instruments, produced the first integrated circuits, or microchips, which package a computer's entire central processing unit (CPU) onto a single chip— solving some problems but introducing others, such as radio frequency interference (RFI).

Microchips may have been that crucial spark that launched the computer age, but computers, in one way or another, have been around since the dawn of humankind. Computers need not be digital, nor electronic. A digital device converts inputs to (usually) binary information, but an analog device can handle a continuum of inputs (e.g., a thermostat, which is non-electronic, or a record player, which is electronic). And mechanical aids predate humankind's harnessing of electricity. Think of the abacus, developed by the ancient Chinese (although the abacus may have originated in the Middle East), which has been around for thousands of years, but is more properly termed a calculating machine than a computer. Other analog mechanical aids to calculation, such as the Antikythera mechanism for astronomy, John Napier's Bones (which were numbering rods; Napier invented logarithms, permitting the simpler operation of the addition of logarithms to take the place of more complex multiplication operations), and the slide rule for multiplication and division (which took advantage of logarithms for quick calculations), were developed in past centuries. Also consider Stonehenge, that arrangement of large stones in England set up thousands of years ago: with the help of the sun, the stones were used to make astronomical predictions. In fact, Stonehenge might be considered the first computer, but not the first calculating device. Simple tallying was facilitated by nothing more than the digits on our hands, which gave way to rocks and sticks and other physical objects when counts needed extending— and which ultimately led to abstracting symbols as stand-ins for those physical objects.

When he was very young, Blaise Pascal, who would become one of the most prolific mathematicians of the Renaissance, invented and patented an adding machine to help his father, a tax collector (the job was called a "tax farmer" then). Pascal was only a teenager when he perfected the adding machine, which he jury-rigged to multiply and divide quantities as well. Another mathematician, Gottfried Wilhelm Leibniz, who co-created calculus along with Isaac Newton, later improved on Pascal's design (as did Samuel Moreland), calling his version the "step reckoner"; but this manual mechanical calculator never achieved widespread popularity largely because of its high production costs.

Logarithmic and trigonometric tables were vital for astronomical, engineering, and mathematical calculations during the Renaissance. Lacking access to operational mechanical calculators, human beings were by and large left to do the arithmetic by hand. These "human calculators" were called "computers"—thus, even the word "computer" predates the electronic computer. *Merriam-Webster's* definition of a *computer* is "one that computes; specifically: a programmable usually electronic device that can store, retrieve, and process data." *One that computes*; for example, during World War II, individuals who tabulated and computed results by hand, such as for ballistics' firing tables, were quite literally referred to as computers; these "computers" were usually women. Such human computers were a critical part of the Manhattan Project, despite the burgeoning availability of electronic computers at the time.

By even the nineteenth century, however, Charles Babbage, a professor of mathematics at Cambridge University, was frustrated with the errors on mathematical tables that human computers, left to their own (lack of) devices, produced. Therefore, he sketched out his Difference Engine, a fully mechanized tabulator that could perform calculations and solve polynomials, and his Analytical Engine, a programmable mechanical computer utilizing a punched card system lifted from the state-of-the-art Jacquard loom mechanical loom (developed by Joseph Marie Jacquard).

Despite receiving funding from the British government to build his computers, which would have weighed in excess of two tons and required thousands of gears and other moving parts, the machines were not constructed, largely due to insufficient tooling technology. (The computers were built in full, but only recently, and are now proudly displayed in the London Science Museum.) But the Analytical Engine's punched-card system of programming and storing information became the standard for electronic computers over a century later.

More interestingly, Babbage had an epistolary collaborator: Ada Lovelace, the daughter of the poet Lord Byron. Lovelace was the world's first computer programmer. She contributed a *Notes* addendum to an article on Babbage's Analytical Engine, describing possible ways of programming the machine, suggesting a program to calculate Bernoulli numbers as well as explicating the first programming loop. Lovelace viewed computer programming as a "poetical science," a sort of nexus between science and art, and she was comfortable traversing both worlds.

It would ultimately fall to the late nineteenth century American inventor Herman Hollerith, who worked for the U.S. Census Bureau, to put the punched-card concept into practice. By the completion of the 1890 census, many millions of punched cards had been used to compile demographic statistics. Hollerith later founded the Tabulating Machine Company, specializing in tabulating equipment, which was eventually consolidated in the 1920s, forming IBM. Hollerith's early punched cards eventually transformed into IBM's 80-column punched cards and were affectionately referred to as Hollerith cards.

So, by the turn of the twentieth century, automated calculating machines were being used by U.S. census takers, but these machines were neither digital nor programmable. In the 1920s, bulky devices called Lehmer sieves could factor simple mathematical expressions; although electronic, they relied on an array of chains, gears, and switches. Prior to World War II, the American engineer Vannevar Bush helped build the differential analyzer, a mechanical analog computer that could solve differential equations. And also before the war, a German engineer named Konrad Zuse constructed a primitive programmable computer called the Z1— in his parents' apartment.

During the Second World War, the electromechanical Harvard Mark I Computer (or Automatic Sequence Controlled Calculator) was assembled at Harvard University by Howard Aiken. Aiken was driven to create an automated calculating machine out of necessity: he needed to work through complex calculations for his doctoral thesis. But he quickly realized that the scope of such a machine could be larger than for just solving math problems. Aiken appealed to IBM for help with the construction, and they acquiesced. In Aiken's proposal to IBM, he justified the need for computational power with an appeal to continued scientific progress:

> The intensive development of the mathematical and physical sciences in recent years has included the definition of many new and useful functions nearly all of which are defined by infinite series or other infinite processes. Most of these are inadequately tabulated and their application to scientific problems is thereby retarded.
>
> The increased accuracy of physical measurement has made necessary more accurate computation in physical theory, and experience has shown that small differences between computed theoretical and experimental results may lead to the discovery of a new physical effect, sometimes of the greatest scientific and industrial importance.[*]

Besides merely arithmetic operations, Aiken proposed that the machine be able to handle signed numbers, grouping symbols, powers, logarithms, different bases, trigonometric functions (like sine, cosine, and tangent), hyperbolic functions, and superior transcendental, as well as being capable of computing finite and infinite series, finding real-valued roots of equations, calculating solutions to differential equations, and numerically integrating and differentiating functions.

Before the Mark I was finished, Aiken was called to serve the Allies during the war; but by 1944 the now-completed Mark I was Aiken's responsibility again, courtesy of the Navy. Assisting him with the computer was Grace Hopper, who had earned a Ph.D. from Yale in mathematics a decade before; Hopper would eventually reach the rank of admiral. When programming the Mark I, Hopper removed a dead moth that was blocking the computer tape and a relay— thus, as she put it, "debugging" the program.

The Mark I was used to solve war-related problems, such as those relating to radar design and, secretly, atomic bomb calculations. By 1945, Aiken built a second computer, the Mark II, at the request of the Navy. After the war the Mark III and IV were constructed. But the machines were slow. The Mark I was originally designed to use vacuum tubes, but, because of cost constraints, was bogged down by many moving parts, such as switches, punched tape (used to feed programs into the machine), mechanical relays (which were like on-off switches, but ones directly controlled by currents of electricity), and adding accumulators. Moving parts bend, break, and break down, and were thus quickly found to be unreliable in the long term.

---

[*] http://history-computer.com/Library/AikenProposal.pdf

The shift from the electromechanical, like the Mark I, to the purely electronic was facilitated by the mathematicians John von Neumann (arriving at the concept of a stored-program digital computer, now called a von Neumann architecture computer, where both programs and data are treated uniformly in memory— a structure which Aiken resisted)* and Alan Turning (who wrote a paper in the mid-1930s that laid the foundation for Universal Turing machines, or machines which could compute by executing stored instructions), who both paved the way for digital, electronic, programmable computers like the Colossus series (designed by British engineer Tommy Flowers and used for cryptanalysis at Bletchley Park during World War II), the reprogrammable ENIAC (Electronic Numerical Integrator and Computer, built in the late 1940s by John Mauchly and J. Presper Eckert), and their host of variants like the EDVAC, which are the direct ancestors to the personal computers we use today.† Those first machines worked in decimal, not binary, and were electronic multi-room-sized monstrosities of wires and searing-hot vacuum tubes which, although much faster than electromechanical relays, quickly burned out, sometimes interrupting complex calculations in their smoldering tracks. Vacuum tubes were eventually put to pasture by the transistor.

Transistors figured into that first phase of materials Bell Labs sent to schools in the sixties, but computers did not. By the end of the decade, however, many high school students received their first exposure to computers courtesy of the "Understanding Computers" kit. Inside was a strange looking Bell Labs device constructed entirely out of paper and die-cut cardboard: the CARDIAC, or **CARD**board **I**llustrative **A**id to **C**omputation, created by the visionary Bell mathematician David W. Hagelbarger. The single-address, single-accumulator-based CARDIAC, which was a teaching tool about computers, not an actual computer, needed (rather fittingly) just a single power source to run programs on its hardware: you. As the manual, packaged with the device, explains:

> [Y]ou will be the energy source. *You* will operate the slides and transfer data from one section of CARDIAC to another. You will even do the arithmetic that must be done in the accumulator. This in no way detracts from CARDIAC's power as a learning tool. Remember, you are not working with CARDIAC to learn arithmetic, but to learn how a computer operates.

---

* The Harvard Mark I computer did not have a von Neumann architecture, because program instructions and data were treated differently in memory.

† The UNIVAC was the first electronic computer used for the census. CBS later featured the UNIVAC on-air to help project the results of the 1952 presidential election.

The device is hand-operated, no electricity required. A dynamic flowchart that relays operation instructions rests on the CARDIAC's front face: move slides, copy contents of cells, advance cards, test the accumulator. The Instruction Decoder, under which is a small viewing slot, is especially helpful in this regard, translating the CARDIAC's technical instruction set into simple steps for you to follow.

"Cardboard 'Computer' Helps Students," a 1969 flyer promoting the "small, hand-operated, cardboard model computer" called CARDIAC, the fifth of the Bell System kits distributed to high school students, boasts that

> CARDIAC can be assembled in minutes. It has most of the equivalent parts of larger, digital computers— accumulator, instruction register, memory cells, and input/output system— and a repertoire of ten instructions, enabling it to solve some surprisingly difficult problems... [since it is] [d]esigned to illustrate the operations of a computer and serve as an introduction to programming... [while leading] the students through ten programs, ranging from simple addition to complex computer game playing.

The flyer also lists the materials contained in the kit: besides the die-cut sheets, slides, and five "bugs"— or memory cell placeholders— four of which were spares, and the CARDIAC manual, the computer kit came with a short film called "The Thinking ??? Machines" that was "designed to create classroom interest and discussion," a Vu-Graph CARDIAC (an abbreviated version of the CARDIAC constructed of overhead transparencies, made expressly for teacher-led programming demonstrations), and five silent super 8 films. As the sixties gave way to the seventies, Hagelbarger's creation would become a familiar sight in mathematics and science classrooms across the country.

 Like Brattain and Bardeen, David Hagelbarger was one of Bell Labs' post-World War II top-talent young hires. Born in 1921, he grew up in Ohio, earning a bachelor's degree from the small liberal arts institution Hiram College in 1942, and a doctorate from Caltech in 1947 under the tutelage of the experimental physicist Robert Millikan. In 1949, he received an offer letter from Bell Labs— which represented quite a change from his years of lecturing in aeronautical engineering at the University of Michigan. He took the job.

The rail-thin Hagelbarger was methodical, good with his hands, and very risk-adverse. Bell Labs' engineers were required to wear ties, but he feared sustaining an injury— or worse— in his beloved machine shop. So, instead of wearing neckties that might inadvertently wrap around a drill press and suck him in, he wore bow ties.

In 1954, after an eventful weekend in which an experiment went bad, Hagelbarger searched for something new to study. Inspired by an issue of the magazine *Astounding Science Fiction*, Hagelbarger decided to build an "outguessing machine," a machine that could effectively read minds. Specifically, this mind-reading machine played an electronic version of the old game "matching pennies," also sometimes called "odds or evens." Matching pennies involves two players clandestinely placing a coin, heads up or heads down, on their palms, and then closing their hands in a tight fist. Both players then open their palms simultaneously for each other to see. If the coins match, one of the players wins; if they don't, the other player wins.

The basic idea of the Hagelbarger matching-pennies machine is as follows: you choose one of two options, + or -. The machine then makes a prediction. If it guesses your choice correctly, you lose the round; if not, you win. At first, the chances of the machine guessing correctly are no better than a coin flip. But, after a handful of rounds, the machine detects patterns— no matter how randomly you try to mix up your choices— and thereby becomes fairly adept at predicting the future based on past outcomes.

In a dramatic flourish, Hagelbarger affixed two rows of lights at the top of his outguessing machine. Each column of lights served as a placeholder for a round of man-versus-machine; if you could light up your entire row of lights before the machine lit its row up, you won the game. Hagelbarger wanted to show, not just to tell; he wanted people to see, not just to imagine.

Hagelbarger published a paper[*] describing the mechanics of his outguessing machine and its implications for the telephone industry. Originally he

---

[*] http://seed.ucsd.edu/~mindreader/SEER.pdf

wanted to have the words "outguessing machine" in the title of the paper, but AT&T urged him to pen something more serious; hence, the SEER, or SEquence Extrapolating Robot, was born.

Hagelbarger begins his paper on the SEER by explaining how playing the penny matching game randomly isn't good strategy, since psychology, as much as probability, is involved: "If you are smarter than your opponent, you should be able to guess which way he will choose and make your own choice so as to beat him more than half the time." Furthering the point, Hagelbarger references an 1845 short story by Edgar Allen Poe entitled "The Purloined Letter," one of the first detective stories ever published. The plot centers on the search for a missing letter; the detective, Dupin, uses the odds and evens game to make a comment about strategy.

> "Yes," said Dupin. "The measures adopted were not only the best of their kind, but carried out to absolute perfection. Had the letter been deposited within the range of their search, these fellows would, beyond a question, have found it."
>
> I merely laughed— but he seemed quite serious in all that he said.
>
> "The measures, then," he continued, "were good in their kind, and well executed; their defect lay in their being inapplicable to the case, and to the man. A certain set of highly ingenious resources are, with the Prefect, a sort of Procrustean bed, to which he forcibly adapts his designs. But he perpetually errs by being too deep or too shallow, for the matter in hand; and many a schoolboy is a better reasoner than he. I knew one about eight years of age, whose success at guessing in the game of 'even and odd' attracted universal admiration. This game is simple, and is played with marbles. One player holds in his hand a number of these toys, and demands of another whether that number is even or odd. If the guess is right, the guesser wins one; if wrong, he loses one. The boy to whom I allude won all the marbles of the school. Of course he had some principle of guessing; and this lay in mere observation and admeasurement of the astuteness of his opponents. For example, an arrant simpleton is his opponent, and, holding up his closed hand, asks, 'are they even or odd?' Our schoolboy replies, 'odd,' and loses; but upon the second trial he wins, for he then says to himself, the simpleton had them even upon the first trial, and his amount of cunning is just sufficient to make him have them odd upon the second; I will therefore guess odd'; — he guesses odd, and wins. Now, with a simpleton a degree above the first, he would have reasoned thus: 'This fellow finds that in the first instance I guessed odd, and, in the second, he will propose to himself upon the first impulse, a simple variation from even to odd, as did the first simpleton; but then a second thought will suggest that this is too simple a variation, and finally he will decide upon putting it even as before. I will therefore guess even' guesses even, and wins. Now this mode of reasoning in the schoolboy, whom his fellows termed "lucky," — what, in its last analysis, is it?"

"Rather than playing it safe," Hagelbarger explains, "the machine tries to outwit his opponent and thereby win more than half the time." Two assumptions were built into the machine: (1) People won't be able to play truly randomly, and when they fall into the trap of consistency, the machine will win; and (2) If the machine is winning, its moves should be correlated with the player's, but if it's losing, its moves should be completely random. For the machine to play strategically, eight states of memory, each requiring its own memory register, were built into the device. These "states of play" had to account for three key questions: Did the machine just win or lose? Had the machine won in the game prior? and Did the machine play in the same way, or a different way, as in the previous game?

The outguessing machine's results were astounding, winning against Bell Labs employees significantly more than half the time— the odds of which, by mere chance alone, were about 10 billion to one.

But was Hagelbarger's outguessing machine optimal? Claude Shannon, who worked at Bell Labs and was friendly with Hagelbarger, didn't think so; Shannon believed he could make a simpler yet more effective outguessing machine than Hagelbarger's.

Claude Shannon was born in Petoskey, Michigan, graduating from Gaylord High School. He excelled at science and mathematics, and he had a knack for making working mechanical and electronic machines. After graduating from the University of Michigan, with degrees in mathematics and electrical engineering, Shannon was admitted to MIT, where he was not only an electrical engineering student but also Vannevar Bush's laboratory assistant, assisting Bush with his differential analyzer. In 1938, Shannon connected Boolean algebra, conjured in the nineteenth century by the mathematician George Boole— who linked logic to mathematics, in part by using a binary approach— to electrical switching circuits; specifically, Shannon's master's thesis, "A Symbolic Analysis of Relay and Switching Circuits," was the first paper to make the connection between Boolean algebra and logic gates, which block or permit the flow of electrical current. In 1940, Shannon received his Ph.D. from MIT.

After joining Bell Labs during World War II, Shannon was put to work in cryptography. Several years after the war ended, when he was only 32 years old, Shannon published a revolutionary two-part paper entitled "A Mathematical Theory of Communication" in the *Bell System Technical Journal*. All information could be encoded in the same way, he claimed, using binary digits, or, in a term that quickly caught on, bits, thereby eliminating errors in transmission.

Though shy, he had a playful side. Shannon enjoyed riding a unicycle through the Bell Labs hallways; he often juggled while he rode. He also constructed machines that did strange things, such as the THROBAC, a calculator that worked with Roman numerals (it was a THrifty ROman numeral BAckward-looking Computer), the mechanical maze-running mouse Theseus, and the Ultimate Machine: a small box with a switch that, when flipped on, would produce a hand from the inside box that simply flipped the switch back off and then retracted back into the box. The Ultimate Machine was either the most brilliant postmodern electronic device ever conceived, or the silliest.

At Bell Labs, Shannon enjoyed hanging around with Hagelbarger, who also had prodigious mechanical skills; they often lunched together in Hagelbarger's lab, where Shannon first encountered the outguessing machine. Shannon built his own, ultimately simpler, version, using fewer relays and breaking the game down into fewer possible situations than Hagelbarger's. The two outguessing machines, Hagelbarger's and Shannon's, were pitted against each other, with an "umpire machine" officiating the contest. As Hagelbarger conceded in his paper on the outguessing machine, Shannon's won by a stunningly lopsided ratio of 55 to 45.

A decade after his machine was defeated by Shannon's, Hagelbarger began working on the CARDIAC. His idea for the CARDIAC was simple— show, not just tell, the story of how a computer works, complete with inputs, outputs, counters, and memory cells (akin to the status lights on his outguessing machine, which visually relayed the contest of man-versus-machine)— but it was hardly self-evident. In the sixties, high schools had neither the budget, nor the resources, nor the facilities, nor the trained personnel to house a fully working computer. Computers then were electronic digital leviathans— still relatively new on the scene, evolving rapidly, but mostly a mystery— in addition to being mostly the purview of large organizations: the government, the military, massive corporations, academic institutions, and hybrids of these organizations, such as Bell Labs. The very specialized, technical knowledge necessary to run a computer, let alone understand how one worked, wasn't readily available; in fact, though awed by the sketchiest outlines of computers that they had seen in science fiction films and television shows or read about in books and pop magazines, most Americans couldn't quite fathom the purpose of computers, besides for arithmetic calculations, if they even realized that much. Only a handful of years earlier Thomas Watson, the president of the International Business Machines Company (IBM), spoke for many when he said, "There is a world market for maybe five computers."[*] And Howard Aiken is attributed a similar quote from 1947: "Only six electronic digital

---

[*] Though this quote may be apocryphal.

computers would be required to satisfy the computing needs of the entire United States." Life seemed to be running smoothly and progressing quite nicely without computers: massive cars prowled streets, extra-deluxe appliances filled homes, and surfeits of goods stocked store shelves, all without a computer in sight. Analog ruled the roost. If all the computers in the world had suddenly stopped working in 1965, what percentage of the world's population would have been affected? What percentage would have even noticed their absence?

As rock and roll was finding its grove, there was also a computer revolution afoot, but it was behind the scenes, almost imperceptible. In his paper on the outguessing machine, Hagelbarger set the stage for the future of computers: machine learning. "As machines get more and more complicated, it seems likely that not only routine matters but some things which we now call thinking will be done by machine.... Perhaps in an extremely complicated situation it might be easier to design a machine which learns to be efficient than to design an efficient machine as such." Hagelbarger thought a lot about whether machines could think. In 1968, with Saul Fingerman, a Bell Labs PR hand, Columbia University graduate, and author of many of the Bell Labs science materials, taking the lead, the film "The Thinking ??? Machines"[*] was released.[†]

"The Thinking ??? Machines" was packaged with the CARDIAC kits sent to schools. Unlike "The Genesis of the Transistor," which not only demonstrates the technology but outlines the history of its discovery, "The Thinking ??? Machines" runs like a protracted thought experiment on phenomenology, the genesis of which is likely a footnote in Hagelbarger's paper contemplating the nature of thinking. The footnote reads as follows: "Some people prefer to stop calling it 'thinking' as soon as a machine does it. Most of us agree that it is a higher form of intelligence for Newton (or Leibnitz) to invent The Calculus than it is for a school boy to learn it, but is the school boy thinking?"

In the film, after setting the stage with a visual montage of science fiction computers, including the Robot from the popular Irwin Allen television series *Lost in Space*, the key question is posed: Do computers think? A generic-looking robot (or, as the narrator clumsily calls it, "row-bit") delves into its "Data Bank Dictionary" to pull out various meanings of the words "to think," each of which is given a thorough treatment. "To think" could mean:

- *To call to mind, to remember.* Storing and retrieving information might not be a particular strength of human beings, but computers excel at it. "Once programmed," the narrator explains, "it [a computer]

---

[*] https://www.youtube.com/watch?v-clud9H8DXU

[†] The film's title was surely influenced by Shannon's paper describing his version of the outguessing machine, which was called "The Mind-Reading (?) Machine."

can refer to its own memory for instructions and data." This basic notion is also one of the key conceits of the CARDIAC.

- *To subject to the process of logical thought.* Logic "and" and "or" circuits are demonstrated; logical "thinking" is another of a computer's strengths, since computers are in essence highly complex networks of logic. Although much of the public first learned how effective computers can be at the logical "thought" vital to playing chess after Garry Kasparov lost to IBM's Deep Blue supercomputer three decades later (in 1997), the film shows a Bell Labs scientist competing against a room-sized computer, which "plays a respectable game of chess"; a cathode ray tube (CRT) monitor even graphically plots the layout of all the pieces on the chessboard. Interestingly, the movie *2001: A Space Odyssey*, which was also released in 1968, features a similar, albeit fictional, chess match between a human (David Bowman) and a computer (HAL 9000).

- *To form a mental picture of.* In a display of stereotypically sexist behavior, we see two boys, one of whom is smiling with eyebrows furrowed and has a thought bubble above his head filled by a caricature of a buxom blonde. The other boy has just mouthed "38-24-36"—the blonde's dimensions. The numbers have been translated into a mental image. A robot approaches them, but can't understand their conversation. Robots, who "aren't nearly as imaginative," also aren't nearly as sexist.[*] But computers can help engineers simulate circuitry design and physicists visually plot satellite paths on CRTs. The ability to form images, if not to form *mental pictures* of images, is a great strength of computers.

- *To perceive or recognize.* Here's where computers (still) fall short. A computer has trouble coming to grips with the simple fact that all of the following symbols refer to the same letter (and the same ASCII, or

---

[*] When we examine the CARDIAC manual later on, we'll encounter several more examples of overt sexism. To be fair, though, there are quite a few professional women shown operating and perhaps programming computers in "The Thinking ??? Machines."

American Standard Code for Information Interchange, code): A, *A*, A, **A**. Language translation isn't a cakewalk for computers, either, because there is no one-to-one correspondence between all the words and symbols in different languages, with idiom and usage throwing further complications into the mix. (So much for building a HAL 9000.)

- *To have feeling or consideration for.* Although programmers might fall in love with their computers, the sentiment isn't reciprocated. Computers don't have feelings and thus, luckily, can't play favorites or become bored. They can, however, be made to simulate emotions.
- *To create or devise.* "So far," the narrator tells us, "no computer has ever composed a hit song." But in the mid-sixties the futurist Ray Kurzweil appeared on the CBS game show *I've Got a Secret*. He played a piano composition for a studio audience that— surprise!— was written by a computer,[*] courtesy of a program Kurzweil coded. The song's not too bad: a completely random-sounding string of notes wouldn't be predictable, able to be anticipated note-by-note, by a human user and thus would sound like noise, while a too-predictable melody would sound formulaic and passé; Kurzweil's song is somewhere in between. But was his computer demonstrating *creativity* by composing the piece? That's as likely as an imminent singularity (or any singularity, for that matter).

The question remains: Can computers think? The problem with answering the question is that the definition of "thinking" has multiple meanings. Computers process information in bits. The notion of a bit, or binary digit, is illustrated in the film by means of throwing a light switch: the bulb connected to it will turn on or off. A single bit thus offers the smallest amount of information transfer possible. Running bulbs and switches in parallel increases the number of bits, permitting more complex encoding.

Although not mentioned in the film, the light bulb demonstration calls to mind Bell Labs' research mathematician George Stibitz's use of relays to perform mathematical calculations, courtesy of electrical current, flashlight bulbs, and binary arithmetic. Stibitz's computer, which he built on his kitchen table in 1937, quickly received the derisive nickname the K-Model (Kitchen-Model) computer. Although at first Bell Labs executives weren't impressed with the K-Model, within a year they agreed to build a large-scale version, called the Complex Number Computer, mostly because Bell Labs employees were getting stymied by difficult mathematics problems that required speedier solutions. Later, Stibitz would operate a relay computer from a remote location, paving the way for the internet.

---

[*] Perhaps the first computer that could generate music was the Ferranti Mark I.

The CARDIAC manual singles out Stibitz for his development of the first electrical digital computers, though his relay-based machines were "primitive by comparison" to modern computers. (Although this may be hard to believe, Stibitz is the only individual in the history of computing that the Bell Labs-centric CARDIAC manual mentions by name.) Rather than light bulbs and switches, though, modern computers make use of magnetic encoding via tapes and disks, and are capable of processing millions of calculations per second with mind-blowing accuracy, as the film makes mention.

Although flawed, "The Thinking ??? Machines" successfully shifts the focus of the student who perhaps became disappointed when he or she realized that the CARDIAC wasn't an *actual* computer— how many high school students had even seen a computer?[*]— but a mockup metaphor for a computer, a simulacrum, an aid to understanding computer architecture. But the mockup would help prepare students for when they eventually received their highly prized and strictly rationed time with computers,[†] like batting practice prepares baseball players for in-game action.

Thus, the key theme running through "The Thinking ??? Machines" isn't the importance of investigating whether or not computers can think; rather, it is this: *Understanding how a computer works is a prerequisite for being able to use the machine properly.* Credit for this theme lies at least in part with the technical consultant of the film, Thomas H. Crowley.

Crowley was born in 1924 in Bowling Green, Ohio. After a stint in the army during World War II, he arrived at Murray Hill in the mid-fifties with a newly minted doctorate in mathematics from Ohio State University. His research interests quickly turned to the nascent discipline of computing, not only earning patents and helping to develop UNIX, a programming language, but also finding ways to spread the gospel of computers to the masses. For example, as the author of his obituary[‡] notes,

---

[*] According to the book *How to Build a Working Digital Computer* (1967), by the middle of 1965 there were only 18,000 computer installations in the entire United States, "serving the government, science, schools, and industry."

[†] With batch processing tightly regulating the flow of programs through the machines, programmers had to write their code in a non-dynamic, non-interactive manner: on punched cards or computer tape. The cards or tape were handed to computer operators, who input the programs and, later (minutes? hours? days?) retrieved and delivered computer output. Thus debugging had to be done on paper, with programs being first run in the mind's eye; running a program with a logic or syntax error might cost days or even weeks of a programmer's time before an opportunity to correct the code became possible. Later time-sharing machines allowed programmers more flexibility. As a teenager, Bill Gates would exploit loopholes and rely on serendipity to take advantage of computer time on time-sharing machines and mainframes, helping him become a master coder by the time he enrolled at Harvard. Time-sharing terminals were ultimately supplanted by the microprocessors in personal computers.

[‡] http://obits.nj.com/obituaries/starledger/obituary.aspx?pid=172608782

[Crowley] taught Union County, N.J., adult education classes on how computers work at a time when computers were still an exotic mystery to most people. As an outgrowth of his classes, he wrote an introductory book called *Understanding Computers* published by McGraw-Hill in 1967, which sold well over 100,000 copies, a best-selling technical book. Computers were the focus of his professional career, and he volunteered his expertise to help individuals, schools, and church groups with their computer needs.

It is not clear how much— if at all— Crowley, who died in 2014, contributed to writing the CARDIAC manual.* Only Hagelbarger and Fingerman are credited with authorship. But Crowley's book *Understanding Computers* was packaged with the CARDIAC kits sent to high school teachers, and Crowley's passion for educating laypeople about that which he devoted his life certainly influenced the methodical, colorful, breezy spirit of the manual.

And about that manual: after bragging in the Preface that Bell is one of the largest users of computers in the U.S., and running through a quick history of computation— highlighting some of the same milestones already touched on here, such as the first electrical digital computers, the birth of the transistor, and the creation of information theory— the authors turn toward the future, noting the increased role of computation permeating all of Bell's research. "It is no exaggeration to say the story of Bell Laboratories and computers is a significant one," the manual reads. "As a by-product of this story [of Bell's involvement with computers], CARDIAC was developed, which we hope will help you understand computers. –G.I.R." That's the entire history of the development of the CARDIAC as background in a single sentence, signed with a mysterious set of initials.

But perhaps a clue to the CARDIAC's origins can be found in the "Selected General Bibliography on Computers" at the end of the CARDIAC manual. Hagelbarger lists a dozen books but one in particular stands out: *The Computer Revolution*, Edmund Callis Berkeley's 1962 book.

Edmund Berkeley, born in New York in 1909, earned a degree from Harvard in mathematics and logic. His first profession was an actuary, where he clerked for Mutual Life Insurance and landed a more prestigious role at Prudential Insurance of America. But by the end of World War II, he turned his attention toward computers, having been influenced by visits to Bell Labs, in which he worked with George Stibitz's Complex Number Computer, and Harvard, where he saw the Harvard Mark I; he helped to build the Mark II. Berkeley also believed in abolishing nuclear war— so much so that when Prudential restricted him from dreaming up ways to reduce nuclear weapons even in his spare time, he quit his job to form a consulting company called Berkeley Associates. He also began writing forward-thinking books about

---

* Called *An Instructional Manual for CARDIAC: A Cardboard Illustrative Aid to Computation*, it can be found at https://www.cs.drexel.edu/~bls96/museum/CARDIAC_manual.pdf

computers, such as *Giant Brains, or Machines That Think,* to great acclaim. In *Giant Brains,* Berkeley prognosticated about the future of computers, even sketching out the first personal computer, Simon— "because of its predecessor, Simple Simon [from Mother Goose]," he wrote— which was a "very simple machine that will think." Simon ran on Harvard, rather than von Neumann, architecture, which was no surprise considering Berkeley's close work with the Harvard Mark I and II. As he explains,

> [Simon] is a miniature mechanical brain containing 129 relays, a stepping switch, and a five-hole paper tape feed. It take in numbers and instructions on a punched paper tape, and shows the answers to a problem in lights. It can take in numbers from 1 to 255 in binary notation, and it can perform any of nine operations including addition, subtraction, greater than, selection, etc.

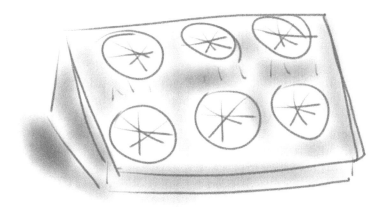

Like a Turing machine, Simon ran programs from paper tape, but the programs were limited to arithmetical and bitwise operations. Simon was eventually constructed and sold hundreds of units in the 1950s.

Here's where our story takes an interesting turn. Besides the Simon, Berkeley went on to design and sell a number of small primitive computers, such as the Geniac (Genius Almost-Automatic Computer; see the sketch above), Tyniac (Tiny Almost-Automatic Computer), and Brainiac (Brain-Imitating Almost-Automatic Computer), among other –iacs. Is it a coincidence that Hagelburger's machine is called the CARDIAC? According to the History of Computers and Computing website,[*] the answer is no. In fact, the website goes out on a limb by pinning down the CARDIAC's development to a man other than Hagelbarger: Irving Becker, who started his career in electronics by making radio kits but by the mid-sixties was producing computers through his company, Comspace Corporation (first named Arkay International). Becker's most famous creation is the Computer Trainer Model 650, also called the CT-650, and also sometimes affectionately referred to as the "paperclip

---

[*] http://history-computer.com/

computer."* He was especially interested in educating students about computers. As the website tells it,

> By the 1960's Irving Becker was developing many educational products, including the digital computer CT-650 and a cardboard kit for Bell Laboratories, called CARDIAC (a reference to its cardboard construction and the names of other kits like the popular Simon, Brainiac and Geniac of Edmund Berkeley).†

An internet posting from 2008, with the headline "New-old stock of Bell Labs's cardboard teaching computer, the CARDIAC,"‡ updates its readers on the status of the "lone remaining dealer of the original Bell Labs Science kits": better act fast to purchase any remaining inventory (including CARDIACs), because the company was slated to go out of business soon. The name of this company? Comspace Corporation. Although their website was taken down years ago, it is still accessible via an archival site.§ On the site, you'll see that their "fantastic product line" included a number of the old Bell Labs' kits, not just the CARDIAC, "under exclusive agreement with BELL LABORATORIES."

So the CARDIAC's origins, which at this point can only be inferred, are as mysterious as the three initials that adorn the manual's Preface: G.I.R. To whom is G.I.R. referring? It's not Hagelbarger or Fingerman or Becker, or some combination of their first and last initials. It also isn't an acronym for the Bell Labs' Educational Programs and Exhibits Department, which, at least "officially," prepared the science kits.

Perhaps the mystery of the origins of the CARDIAC— if not G.I.R.— can be solved by now turning to a computational instructional device which slightly predates the CARDIAC: the Little Man Computer.

Entrepreneur, academic, professor, author— Stuart E. Madnick has done it all. Born in 1944, Madnick might best be described as an MIT lifer. And not just because the decorated septuagenarian has a staggering *five* degrees from the institution— two in electrical engineering, two in computer science, and

---

* The CT-650 earned the nickname the "paperclip computer" because Becker was thought to have taken the CT-650's processor design from the 1967 book *How To Build A Working Digital Computer*, which detailed how to fashion a computer using everyday household items like screws, spools, light bulbs, and, of course, paperclips.

† http://history-computer.com/ModernComputer/Personal/Arkay.html

‡ http://boingboing.net/2008/03/11/newold-stock-of-bell.html

§ https://web.archive.org/web/20080915051507/http://hometown.aol.com/comspace/

one in management— but also because quickly after earning his Ph.D. in computer science, he joined the faculty at MIT and has taught there ever since, earning such titles as Professor of Information Technology at the MIT Sloan School of Management and the Massachusetts Institute of Technology School of Engineering, Director of the MIT Interdisciplinary Consortium for Improving Critical Infrastructure Cybersecurity, Co-director of the Productivity from Information Technology (PROFIT) program at the Sloan School, and John Norris Maguire Professor of Information Technology. He's even the owner of a fourteenth-century English castle, something he purchased on a whim using the profits from the sale of one of the half-dozen tech companies he founded. After buying Langley Castle, the British government officially designated Madnick the Baron of Langley— as if he needed another title.

Madnick's attention might now be directed at how organizations can integrate information and technology effectively, but back in the mid-sixties, when he was an ambitious young MIT student, his focus was on the basics of computer architecture. Working with his thesis advisor, John Donovan, who was only two years younger than Madnick but on the fast track to academic and entrepreneurial success as well, Madnick arrived at a model to simplify computer architecture. Calling it the Little Man Computer (LMC), the concept was so powerful that, as Hugh Osborne and William Yurcik explain in their article "The Educational Range of Visual Simulations of the Little Man Computer Architecture Paradigm," the LMC "teaching tool for computer architecture education...had endured for 40 years [since it] provides important underlying details of computer operation. Stuart Madnick and John Donovan of MIT originally developed this paradigm, where it was taught to all MIT undergraduate [computer science] students during the 1960s." The LMC paradigm has stood the test of time because the underlying architecture of the binary, digital, programmable computer has remained largely intact for decades; in addition, as Larry Brumbaugh and William Yurcik point out in yet another article about the LMC (there are quite a few out there), the LMC's simplicity makes it a highly effective teaching tool.

The central conceit of the Little Man Computer is that there is, quite literally, a little man— a homunculus— running around inside a "mailroom" within a computer, frantically carrying out commands, retrieving papers from input and output baskets, and adjusting the one hundred different mailboxes (each of which can hold a three-digit number)

accordingly. Right now, wading through the technical details of the LMC isn't important; we will get to those later on. Suffice it to say, though, the LMC exposed, and continues to expose, generations of higher education students to the rudiments of computer architecture, through bare paper-and-pencil models and, especially lately, virtual LMCs which have proliferated on the internet in the past two decades.

But a question remains: since the LMC and the CARDIAC use, as we'll discover, very similar approaches to modeling a computer's architecture, and the LMC was utilized to teach students at one of the most prestigious institutions in the country before the CARDIAC arrived on the scene, was Hagelbarger influenced by Madnick's creation or did he independently arrive at a similar instructional tool?

The parallels between the two instructional tools are legion. Like the LMC, the CARDIAC has one hundred memory cells, along with an input card, an output card, instruction register, instruction counter, single accumulator (or calculator), and a simplified instruction set (containing around a dozen machine language instructions called opcodes, or operational codes). Furthermore, both mockup computers run in base 10 (decimal), rather than base 2 (binary).

Unlike in Poe's "The Purloined Letter," this is where the trail goes cold. One the one hand, the LMC predates the CARDIAC by at least a year,[*] and that would certainly be enough time for someone at Bell Labs to get wind of the goings-on in undergraduate computer science classes at MIT, or to hire a lanky computer science kid straight from the university. The early Bell Labs was populated with young men "trained at first-rate graduate schools like MIT and Chicago," as Jon Gertner observes. One of these young men, who earned his doctorate at MIT and was hired by Bell Labs in the early 1940s, returned to his alma mater in the mid-fifties to finish out his career there as a professor: Claude Shannon. But Shannon's connection to Bell Labs didn't quite end when he left the organization, as Gertner explains:

> At first, Shannon visited Murray Hill occasionally. He would come to talk with his old colleagues like [David] Slepian and Hagelbarger about his most recent ideas, and to hear about theirs and perhaps offer a suggestion or two. Shannon and his supervisors at Bell Labs agreed that he

---

[*] The Computer History Museum webpage (computerhistory.org) reports three release dates for the CARDIAC: 1966, 1968, and 1981. The 1968 date corresponds with the production of "The Thinking ??? Machines" and the promotional push to high school teachers. The 1966 date is, presumably, the original date of manufacture, but as a standalone device, not packaged as part of a science kit. As for the last date, the webpage explains, "The 1981 version differs significantly but in principle is a later modification of the CARDIAC computer. The 1981 version is known as the Information Machine." In addition, there was at least one international version released: the French-language Carton Didactique Accessoire (later called the Ordinapoche).

would stay on the Labs payroll as a part-time employee, so he still maintained an office there.

Did any of those conversations between Shannon and Hagelbarger include talk of a Little Man Computer? It's certainly possible.

Also, don't forget that Bell Labs worked hand-in-hand with the government, corporations, and academic institutions— which included MIT. Cross-pollination of ideas was bound to occur. For instance, Dennis Ritchie, the creator of the C programming language, writes of a loose collaboration between Bell Labs at Murray Hill (where both he and Hagelbarger made their professional homes) and MIT in the development of the UNIX operating system: "the implementors of the IO system were at Bell Labs in Murray Hill, while the shell was done at MIT."* When did this occur? In the late 1960s.†

On the other hand, the CARDIAC has a number of differences with the LMC that can't simply be swept under the rug. For instance, Hagelbarger thought to package the device as a self-contained paper-and-cardboard unit, rather than just a conceptual teaching tool that, quite frankly, didn't need to be tethered to anything physical at all.‡ And the similarities may be attributed to coincidence— after all, how many ways are there to simplify for a beginner what was then a relatively new idea: von Neumann architecture? Hagelbarger and Madnick were probably reading the same small set of available computer books; Hagelbarger lists a dozen such books in the CARDIAC manual's bibliography, all of which predate the LMC.

In addition, notwithstanding the History of Computers and Computing website, it's not at all clear what Irving Becker's contributions to the development of the CARDIAC included. Did his company, Comspace, produce copies of the device? Did his company distribute them? Was Becker familiar with the LMC, since he had a prodigious interest in the link between computers and computer education?

We might instead even flirt with the idea of so-called multiple discovery (or simultaneous invention) to explain the similarities between the LMC and

---

* https://www.bell-labs.com/usr/dmr/www/hist.html

† Also, to be completely fair, Madnick was almost certainly influenced by the CARDIAC, since he revised the LMC in 1979 by streamlining the instruction set. Going forward, we will treat this latter version as the standard LMC without further comment.

‡ Although in practice the LMC was sketched out on paper and blackboard (and much later as a host of dynamic virtual simulations, many available to download online), working through programs on the "device" could just as easily be done purely in the province of the mind, à la Ada Lovelace. But that's true of programming on the CARDIAC as well. I've included the LMC in this book to serve as a counterpoint to the CARDIAC, since not more than a hair's breadth separates their respective designs and their histories seem, in retrospect, inextricably intertwined.

the CARDIAC. Like Newton and Leibniz, who discovered calculus nearly simultaneously, or Charles Darwin and Alfred Wallace, who came up with the theory of evolution at more or less the same time, perhaps Madnick and Hagelbarger independently arrived at the same teaching paradigm.

Or maybe there's another explanation— for the genesis of *both* instructional computers. Consider the TUTAC, or TUTorial Automatic Computer. The TUTAC was the centerpiece of a lengthy 1962 book titled *Basic Computer Programming* by Theodore G. Scott. The TUTAC had a more extensive set of opcodes than either the CARDIAC or LMC, with specified digit shifts, copying, and multiple arithmetic operations populating the instruction set. TUTAC also had a large memory, with the capacity of 10,000 memory cells; the paradigm for input/output was punched cards. And Theodore Scott's brainchild was a decimal-based machine, too.[*] Were Madnick and Hagelbarger influenced by the TUTAC? Perhaps the academic and industrial think-tank culture at the time was simply a hotbed for computer instructional models to spawn from the minds of the preternaturally creative.

Regardless, the mystery of creation remains unsolved.

The Little Man Computer was a product of the ivory tower and was thus unsurprisingly geared toward undergraduates. The CARDIAC, which came from a research lab, was sent to high school students. Did any comparable instructional tool originate in the secondary school system? Enter Fred C. Matt, a teacher who split time at the middle school and high school of a small suburban public school district in Southeastern Pennsylvania from the mid-seventies to the mid-nineties.

Born in 1939, Matt was an iconoclastic figure in the district, a well-known resource to students and faculty interested in technology, especially computers. Matt ran a popular extracurricular computer club and taught mathematics and computer science to kids at a time when standalone, cheap, personal computers were only beginning to make their presence felt in public school classrooms. The high school's 1985 yearbook, called the *Elmleo*, sets the scene:

> The computer age was here, and the students here were determined to keep up with it. Enrollment in the computer courses offered in the curriculum was especially significant as a result of the recently imposed credit requirements in computer studies. The computer facility at the high school was comprised of a multitude of Radio Shack Model 4 personal computers situated in two separate classrooms, each equipped with a line printer.

And who was to staff this new computer science "department" at the high school? A caption underneath a picture of a balding man with a shock of

---

[*] https://github.com/mvanmoer/TUTAC

black hair, casually dressed in a white, open-collared shirt, sitting at a computer terminal next to a male student, answers the question:

> Teaching the fundamentals of BASIC [a high-level programming language], Mr. Matt demonstrates a program to [his student]. After teaching at the Middle School for many years and designing the paper computer, Mr. Matt joined the high school math department as a computer teacher.

It had been a long journey for him, ultimately culminating in being rewarded with a chairmanship of the mathematics department at the middle school. Like Hagelbarger, Matt had a passion for electronics, receiving a degrees in electronics and physics from Temple University, along with a master's in math from Penn State— although all this academic success only came after years of struggles with his own early schooling; Matt credits his parents who "made sure that I at least finished high school at a time when I cared less."

Besides computer programming, he taught computer physics and electronics, and was just as interested in the theory of computers as he was in actually constructing the machines. He even worked as a computer technician for IBM. But it was his passion for education that was most important.

Full disclosure: When I was in middle school in the 1980s, Mr. Matt was my teacher. And, apart from his computer programming classes and after-school computer club, school was mind-numbing. Mark Edmundson, when reflecting on his own schooling experiences in the semi-autobiographical *Why Teach? In Defense of a Real Education*, eloquently puts into words why public schools have failed generations of kids.

> Medford High School, whatever its appearances, was not a school. It was a place where you learned to do— or were punished for failing in— a variety of exercises. The content of these exercises didn't matter at all. What mattered was form— repetition and form. You filled in the blanks, conjugated, declined, diagrammed, defined, outlined, summarized, recapitulated, positioned, graphed. It did not matter what: English, geometry, biology, history, all were the same. The process treated your mind as though it was a body part capable of learning a number of protocols, simple choreographies, then repeating, repeating.

By contrast, Mr. Matt's classes involved pure creativity, an admixture of art and science— even a bit of poetical science, as Ada Lovelace termed it. Matt knew that the study of mathematics and computers, if a teacher wasn't careful, could quickly turn into the kind of nonsensical gibberish Captain James T. Kirk spews off to a bewildered guard while explaining the imaginary card game "fizzbin," from the famous *Star Trek* television episode "A Piece of the Action":

Each player gets six cards, except for the player on the dealer's right, who gets seven.... The second card is turned up, except on Tuesdays.... Oh, look what you got, two jacks. You got a half fizzbin already..... [But don't get another jack.] If you got another jack, why, you'd have a "sralk".... You'd be disqualified. You need a king and a deuce, except at night of course, when you'd need a queen and a four.... Oh, look at that. You've got another jack. How lucky you are! How wonderful for you. If you didn't get another jack, if you'd gotten a king, why then you'd get another card, except when it's dark, you'd give it back....

Matt constantly refined his teaching craft to avoid these fizzbin-type explanations, especially when it came to the nascent discipline of computer science. He carried the burden of knowing that, for many of us, he would deliver our first glimpse of the magic behind computers.

Although an overriding interest in computers was hardly new to me when I met Mr. Matt— I had already been programming for years in BASIC on my own Radio Shack Color Computer 2,[*] along with avidly reading periodicals like *Rainbow Magazine* as well as computer programming manuals— Mr. Matt presented the material in a way that inspired me to extend myself and want to learn much more.

One cold day, before the winter break, he showed me the unusual looking Instructo[†]— a computer made of paper and cardboard that he had patented some years before. Although primarily a teaching tool, the Instructo was also fully programmable, sort of in the way that the electronic Radio Shacks in the school's computer lab were. The difference was, he explained to me, the Instructo Paper Computer (IPC) relied on human beings rather than electricity as power sources to operate slides, do arithmetic, flip switches, and produce output. For those

---

[*] A variant of the then-ubiquitous TRS-80s (running on Zilog Z80 microprocessors)— one of a trinity of influential microcomputers released in the late seventies, along with the Apple II and the Commodore PET— which were certainly not "Trash-80s," as some critics labeled them (because their "operating system and external devices did not work smoothly at times," as *CoCo: The Colorful History of Tandy's Underdog Computer* explains), but had much to recommend.

[†] https://web.archive.org/web/20071011043313/http://www.computermuseumgroningen.nl/mcgrawhill/ipc.html

of my generation, who weren't yet born when the CARDIAC was mass produced but came of age when microcomputers proliferated in homes and schools, the Instructo was the paper computer of choice (among at least several hundred students, that is).

The Instructo lit a fire in my imagination, though I didn't really understand much of it at the time. The commands were obscure and low-level, nothing like high-level BASIC I was used to coding, but also not entirely similar to assembly language, either. To program the Instructo, you had to utilize a mix of low- and high-level techniques; sometimes programming it felt like an exercise befitting a Rube Goldberg machine, and running programs on its cardboard hardware (perhaps we should term it card-ware) was laborious. Although vaguely similar to the CARDIAC and LMC in design— and hardwired with more opcodes than the TUTAC— the Instructo really was its own animal, fresh and original, cut out of whole cloth (or cardboard).

It was also immediately out of date. Like the CARDIAC and LMC, the Instructo was designed to be a teaching tool about computers rather than a pure calculating machine or actual computer; after all, the calculating and incrementing and data transfer has to be done by hand. But the CARDIAC and LMC arrived on the scene in the sixties— well in advance of electronic personal computers, despite the CARDIAC manual declaring the 1960s the *Age of the Computer*, complete with a cartoon of a bewildered boy looking off in the distance as a large mainframe computer streaks by so quickly that it leaves ash and dust in its wake. The Instructo was designed when disco was peaking in the late seventies and computer hobbyist groups like the Homebrew Computer Club were flourishing,[*] and it was trademarked and published, through McGraw-Hill,[†] by the early eighties, right around the time when the Apple II and the Radio Shack line of personal computers exploded in popularity. Why buy a paper computer to teach students when you can get ahold of the real thing? The Instructo was an instant anachronism, a curiosity, a product filling a need from a decade past.

---

[*] The Silicon Valley Homebrew Computer Club meetings were attended by such future computer industry luminaries as Steve Jobs and Steve Wozniak. According to *CoCo: The Colorful History of Tandy's Underdog Computer*, "The club's early appeal was the enthusiastic and free exchange of ideas, technology, and information among its talented members." Opposed to this free exchange was Bill Gates of the company Micro-Soft (which, of course, later dropped the dash and became Microsoft), who decried the promotion of what he termed "software piracy" in an open letter addressed to the club; he was angry that an early paper-tape version of the BASIC programming language he and Paul Allen wrote was being copied illegally.

[†] Which, recall, had a history of publishing introductory computer texts, such as *Understanding Computers* by T. H. Crowley. Matt's wife, Helene, helped him to land the McGraw-Hill publishing contract.

On September 9, 1980, the *Sydney Morning Herald*'s Stephanie Blackett eviscerated the quixotic Instructo in a highly critical review. A self-described "computer illiterate," Blackett dismisses the manual's description of computers, writing, "It was all clear, but I didn't grasp it— too many trees, not enough forest," and calls the process of programming "puerile" with too many details and tasks, largely absent the reasoning for them. "I, like the computer, was a robot obeying orders. I did not understand how the program had been constructed." Going on, she complains that she was "bored stiff," finding that "[i]n some programs, I could ignore the computer and find the answer in five seconds using my head and a pencil." It just wasn't fun, the technophobic author lamented. And when the algebra started cropping up— some of Matt's ready-to-run programs had the Instructo plugging and chugging through mathematical formulas— she was flabbergasted. Clearly, this computer tutorial package wasn't geared toward her. Yet Blackett's conclusion makes clear that her criticisms lie more with the manual's presentation than the Instructo itself: "In a classroom with an inspiring teacher, the paper computer has a place. It is a teaching aid, not a teacher."

Blackett's article could have just as easily been a review of the CARDIAC or the LMC, since so many of her points resonated with computer instructional tools in general. "Finding the answer in five seconds" defeated the purpose of running a program— which was to learn about computer architecture and the *process* of programming. Scribbling off a quick arithmetic calculation was beside the point.

By the early nineties the Instructo's trademark had expired; and with Fred Matt's retirement after three decades of teaching, and the proliferation of ever-more-powerful personal computers, it fell into disuse. The Instructo now has a page on the Computer History Museum webpage devoted to it,[*] almost the only digitally permanent testimony to a nearly forgotten concept.

With the benefit of hindsight, it's clear that some technologies were meant to be bridges, mere stopgaps, between an older technology and an emerging one. Think of 8-track tapes, which used magnetic tape but with analog encoding, as linking Vinyl records and CDs; think of those first bulky handheld mobile phones as linking landlines (and cordless phones) and cellular phones; think of steam-powered cars as linking horse-drawn carriages and gasoline-powered vehicles.

Likewise, we can think of the LMC and CARDIAC— and, to a lesser extent, the Instructo— as filling a gap: computers, relatively new on the scene, weren't available for classrooms' full of students to use, due to size, cost, and facilities' constraints. Time on those early bulky machines, for those lucky enough to procure access, was limited and rationed. The instructional tools

---

[*] http://www.computerhistory.org/collections/catalog/X1015.89

modeled stripped-down, simplified versions of the real thing. The Instructo, like the CARDIAC and LMC, is hardly rare breed, belonging to a class of antediluvian ideas whose time seems to have passed. Which would be true, of course, if computers had changed radically or fundamentally since the 1960s.

Although there have been a host of changes, a computer's basics are still very much intact. In fact, the LMC still features prominently in a perennially popular academic book on computer architecture and organization: *The Architecture of Computer Hardware, Systems Software, and Networking: An Information Technology Approach* by an old hand at computer design, Irv Englander. In fact, a whole chapter— Chapter 6— is devoted to the Little Man Computer. This chapter did as much as anything to publicize the LMC beyond the environs of Cambridge. As Englander explains, after von Neumann laid out the guidelines for computer hardware in the late 1940s,

> [T]he von Neumann architecture continues to be the standard architecture for computers; no other architecture has had any commercial success to date. It is significant that, in a field where technological change occurs almost overnight, the architecture of computers is virtually unchanged since 1951.

After defining von Neumann architecture, which we will also do in more detail in subsequent pages, Englander writes, "[Y]ou will observe that the Little Man Computer is an example of a von Neumann architecture... [since] there is no differentiation between [data and program instructions] except in the context of the particular operation taking place," the key operational component of a computer with von Neumann architecture. "It is the strength of the [LMC] that it operates so similarly to a real computer that it is still an accurate representation of the way that computers work thirty-five years after its introduction." (By the way, Englander received his Ph.D. in the late seventies from— you guessed it— MIT.)[*] And if the LMC still accurately reflects contemporary computer architecture, then so does the CARDIAC— and, to a larger extent, the Instructo, which has multiple registers and other accoutrements that add to its realism at the expense of increased complexity.

Beyond their usefulness at modeling computers,[†] programming on these paper computers provides a level of challenge and a measure of satisfaction

---

[*] When the LMC is mentioned, online or in print, the source is inevitably Englander, not Madnick, since Madnick never published anything on the LMC. For example, two factoids that have proliferated about the LMC— that it was developed in 1965, and that its instruction set was slightly modified in 1979— come from Englander's *The Architecture of Computer Hardware, Systems Software, and Networking*.

[†] Thanks in part to Englander's texts, as well as a proliferation of online emulators, the anthropomorphic LMC model still has pride of place in computer science programs across the country.

that makes the whole enterprise quite worthwhile— though they do dispense with one quality that makes modern computers indispensable. "The truth is, computers are dummies. Really, really *really* stupid," explains Daniel Appleman in *How Computer Programming Works*. Computers don't think, contrary to what Bell Labs' filmmakers might have you believe; rather, Appleman continues, "The smartest computer ever made does not even have the intelligence of a newt.... The power of computers comes not from intelligence, but from speed." When running a paper computer program, we'll have to make do without speed. And without mathematical calculations being automatically completed by the machine.

Yet these pioneers of the paper computer sacrificed speed and automated calculations not only out of necessity but for a key educative reason: when learning about computers, it's important to constrict the student's field of vision as much as possible so as to take in the critical processes and tasks at hand. And these physical devices made of paper and cardboard, these "pulp devices," accomplish that field-of-vision constriction by means of what author Matthew Crawford terms a "jig" in *The World Beyond Your Head: On Becoming an Individual in an Age of Distraction*. Master chefs and others engaging in demanding activities requiring many simultaneous skill sets physically orient and reorient space in clever ways to keep their executive-functioning calls to a minimum; this is called jigging the environment. Likewise, by laying out the physical space of the necessary computer-synthesizing components— such as accumulators, registers, program counters, and the like— in such a way, the "physical jig [of the paper computer's design] reduces the physical degrees of freedom a person must contend with," allowing "one... to keep action on track, according to some guiding purpose,... [and thus] keep attention properly directed." The arrangement of the environment, the jig, when learning about how computers operate and programs run, is made manifest by the paper computer. We might even consider the paper computer to be the jig's apotheosis.

This guide is divided into two parts. The first part explores the basics of computers and programming, and wades through the forests and trees of paper computer design and coding; the second part presents a series of detailed emulation options.

Specifically, in the first part of this guide, using opcodes and mnemonics, we will learn how code for the LMC, CARDIAC, and Instructo in their respective machine languages (decimal-based, rather than binary). Then, we will program all three machines using assembly language routines. Through it all, you will be given a menu of options to run programs, expounded in the second part of this guide: Procure an old device and run programs on the original hardware? Construct a replica? Build the compatible-with-anything Pink Paper Computer, the blueprints of which are presented later on? Use one of this book's numerous virtual paper computers, each courtesy of a Microsoft Excel Macro written in VBA (Visual Basic for Applications)?[*] Download one of a number of CARDIAC or LMC emulators freely available on the internet, found easily after a quick web search? Or simply run the programs in your mind's eye? The choice will be yours.

Before we can run any programs, however, we need to trace an important arc of history— namely, how modern computers came to be.

---

[*] In which the tediousness and self-consciousness of pushing and pulling paper through cardboard is eliminated in favor of automated electronics— ironic, since the supposed appeal of the paper computers was to be their independence from electronics.

PART A

computers and their instructional paradigms

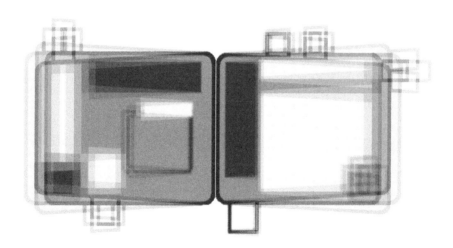

SECTION 1.

the basics of computers

> ↳ Run Turing Machines Programs by Using
> Turing Machines Emulator (see Section 9)
> Emulators Available Online
> Or Your Mind

The story of the paper computer parallels the development of the modern computer, albeit writ small. Thus, before we delve into the particulars of the CARDIAC, LMC, and Instructo, we need trace the story of how the electronic, digital, binary computer came to be.

The development of the modern computer was not a one-person effort; rather, it was the result of the hard work and stunning ingenuity of a vast array of individuals over a number of decades. Four of these individuals deserve special recognition upfront for their contributions: Charles Babbage, Ada Lovelace, Alan Turning, and John von Neumann.

a short biography of charles babbage

Charles Babbage was born in London in 1791, the son of a banker. Early on he had a proclivity for mathematics, learning much of the discipline on his own. By his twenties Babbage had found work as a mathematician, was elected a Fellow of the Royal Society, and began work on his Difference Engine, a complex mechanism designed to perform mathematical calculations. The Difference Engine was merely a warmup, though: Babbage set his sights on creating a mechanical device, called the Analytical Engine, that would perform *any* calculation, even symbolically-based. Furthermore, the Analytical Engine was to be programmable. At first, the British government funded his endeavor; then, when it appeared that construction of the machine was impossible, defunded it.

Although Babbage went on to help found the Royal Statistical Society, and taught mathematics at Cambridge under a prestigious endowed chair, he never got over the British government's slight of his mechanical engines. Babbage died in 1871.

a short biography of ada lovelace

Ada Lovelace, daughter of the poet Lord George Gordon Byron, was born Augusta Ada Byron, the Countess of Lovelace, in London in 1815. Like Babbage, she also showed an early interest in mathematics. Her mother, Lady Anne Isabella Milbanke Byron— who left her husband Lord Byron soon after Ada was born— had Ada tutored in mathematics and science, not because she believed in her daughter's abilities in the subjects, but because she wanted to help mold Ada's temperament through rigorous study. Lady Byron hoped her daughter wouldn't resemble her father, in all senses of the term.

When Ada was just seventeen years of age, she met and began a fruitful correspondence with a middle-aged Charles Babbage. Babbage connected her with Augustus de Morgan, a British mathematician and professor; Lovelace became his student. But Lovelace was especially captivated by Babbage's devices. After being asked to translate an article on the Analytical Engine from Italian to English, Lovelace contributed a long addendum called the *Notes*, in which she explained in detail how to program the device, as well as dreaming up abstract problems for the Analytical Engine, and machines like it, to someday solve, by recognizing the independence of mathematical operations from the numbers used by them:

> In studying the action of the Analytical Engine, we find that the peculiar and independent nature of the considerations which in all mathematical analysis belong to *operations*, as distinguished from *the objects operated upon* and from the *results* of the operations performed upon those objects, is very strikingly defined and separated....
>
> [I]t might act upon other things besides *number*, were objects found whose mutual fundamental relations could be expressed by those of the abstract science of operations, and which should be also susceptible of adaptations to the action of the operating notation and mechanism of the engine. Supposing, for instance, that the fundamental relations of pitched sounds in the science of harmony and of musical composition were susceptible of such expression and adaptations, the engine might compose elaborate and scientific pieces of music of any degree of complexity or extent.[*]

---

[*] https://www.fourmilab.ch/babbage/sketch.html

The Analytical Engine was never built because the tolerances allowed by machine tools at the time were too great, making impossible the construction of precision parts that the massive mechanical contraption required.

Ada's few remaining years were spent devising mathematical strategies for gambling. She died at the young age of 36 and was buried in a grave next to her famous father.

alan turing and his turing machines

Like Babbage and Lovelace, Alan Turning was born in London, but a century later, in 1912. He also showed an early facility with mathematics, eventually enrolling at the University of Cambridge. After arriving at a unique proof of the central limit theorem, a key statistical concept describing how the means of independent random samples converge toward normality, he was elected a fellow and, shortly thereafter, began work on his doctorate in mathematics from Princeton University under his advisor, the mathematician Alonzo Church. He also studied cryptology at Princeton. When he returned to Cambridge, he was tapped to work at Bletchley Park to help decode the German Enigma machine during World War II.* He later went on to formulate groundbreaking advances in the theories of computing, including in the field of artificial intelligence (courtesy of his "Turing Test," which addressed a profound question: How can we know if machines think?).

But Turing's most significant contribution to modern computing lies in his concept of the Universal Turing machine, formulated in the 1936 paper "On Computable Numbers, with an Application to the *Entscheidungsproblem*." In it, Turing discusses computable and non-computable numbers, an open problem formulated by mathematician David Hilbert, the halting problem, and Turing machines. Needless to say, there is quite a bit to unpack in his paper.

To understand the distinction between computable and non-computable numbers, we need to travel back more than a century. David Hilbert, perhaps the preeminent mathematician of his time, presented a list of twenty-three open mathematics problems in Paris at the International Congress of Math-

---

* The biopic *The Imitation Game* (2014) focuses on Turing's life during this consequential period.

ematicians conference in the summer of 1900. Mathematics was—and still is—divided into a number of schools of thought; each school assumes a different ontological approach toward the discipline. Hilbert was a proponent of a school called Formalism, which posits that mathematical symbols and concepts don't need to be tied to anything in reality; math begins and ends with the rules of symbolic manipulation, and the process of manipulation can be thought of as an internally consistent game that need not have any external meaning.

Hilbert was concerned about cracks in the foundations of mathematics resulting from paradoxes such as *Does the set of all sets which are not members of themselves contain itself?* He believed that all formal mathematical systems could be made to be consistent and have a complete set of axioms, and initiated a program, called Hilbert's Program, to help lay the foundations of mathematics onto secure footing though rigorous formalization.

But most of Hilbert's Program came crashing down in 1931, thanks to Kurt Gödel and his incompleteness theorems. Gödel proved that a mathematical system cannot be both consistent and complete—things will always remain unprovable within the system itself.

Gödel didn't completely destroy Hilbert's Program, but Alonzo Church and Alan Turing did. In 1936 they provided an answer to Hilbert's *Entscheidungsproblem*, or decision problem, which asked: Is there some algorithm that can determine if a mathematical statement is true?* Alonzo Church, using lambda calculus, and Alan Turing, using a theoretical construct called a Turing machine, found that there can be no such algorithm.

But what does any of this have to do with computable and non-computable numbers? In Turing's paper, he defines a Turing machine, a finite-state machine (meaning it has a countable number of possible states, and can only be in one state at a time), which has been fed a windingly long paper tape (think of blank cash register tape) beneath a read/write head

---

* Here's how Hilbert phrased it: "The *Entscheidungsproblem* is solved when one knows a procedure by which one can decide in a finite number of operations whether a given logical expression is generally valid or is satisfiable."

that can read off of or write onto the paper tape. The tape is divided into squares; some squares have a single symbol printed on them, such as 0 or 1, which store the inputs and outputs of computations; other squares are left blank. Depending on its current state and input, the Turing machine can erase and write overtop of a symbol, move the tape left or right one square or remain in the same square, change its state, halt, or some combination thereof, as specified by a set of state instructions (i.e., a program). The paper tape can be thought of as infinitely long— or at least as long as it needs to be to run a program through to completion.

Turing machines completely model the universe of possibilities and limitations of all computers (including paper computers), since they capture, in a mathematically idealized fashion, that which is computable using a finite-state machine; essentially, the operations of every computer can be reduced to at least one Turing machine,[*] and a computer or programming language is termed Turing Complete if it can, in theory, do everything that a Turing machine can do.

Turing machines are programmed by specifying the action the machine should take in each state. Starting from some initial, or starting, state, perhaps the Turing machine will move the tape to the right, or print a symbol (depending on some input), or even halt the machine— or perhaps the program slips into an infinite loop, shuttling the machine from one state to the next and back again. If the machine has halted, the pattern of 0s and 1s remaining on the tape will, it is hoped, reveal the answer to the problem.

Let's examine several examples. Consider the following state table:

Turing Machine Program No. 1-1: An Infinite Loop

STATE	READ	WRITE	NEXT MOVE	NEXT STATE
1	Blank	Blank	None	1
1	0	1	None	1
1	1	0	None	1

This first Turing machine program will run forever, no matter the input. For example, if the read/write head encounters a blank square on the tape, it is left alone; the read/write head doesn't move the tape left or right; and the machine resets to the initial state— only to work through the instructions again. If the read/write head instead encounters a 0 as input, it erases it and writes a 1 in its place, doesn't move the tape, and resets back to the initial state. Instead, if the read/write head reads a 1, it changes it to a 0, and then the machine resets to the initial state. A flowchart, or diagram enumerating the sequence of procedures and operations of the program, will make the instruction set clearer. Note that input/output operations on flowcharts are usually drawn as parallelograms.

---

[*] Imagine the futile instructions for assembling a paper computer Turing machine. First, obtain an infinitely long strip of paper. Then, ...

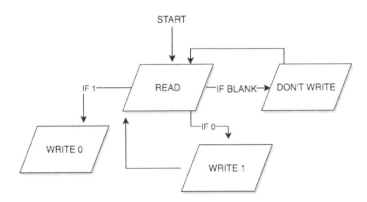

Fig. No. 1. Flowchart of Turing Program No. 1-1.

The program might also be made clearer by looking at the paper tape output. Examine the graphics below, which were created with the help of a Microsoft Excel macro running a Turing machine emulator (the detailed set up and code for which can be found in Section 9 of this guide); the caret (^) symbol relays the position of the read/write head:

B	B	B	B	0	0	1	0	0	1	B	B	B	B	B
				^										

B	B	B	B	1	0	1	0	0	1	B	B	B	B	B
				^										

Now consider this second example, which converts 1s to 0s or 0s to 1s, stopping (or halting) only when a blank space is encountered.*

Turing Machine Program No. 1-2: A Bit Inverter

STATE	READ	WRITE	NEXT MOVE	NEXT STATE
1	Blank	Blank	None	HALT
1	0	1	R	1
1	1	0	R	1

Here, despite having only one state— the initial state— this instruction set performs a useful function: it inverts bits, turning 0 to 1 or 1 to 0 (given an input string), and stopping when encountering the first blank (presumably, the end of the input). Notice that after writing a 0 or a 1, the read/write head moves one square to the right, queuing up for an input on the next square. The flowchart below lays out the instruction set; notice that the ellipse represents a halt (or program stop), while the rectangles stand in for the process of moving the tape left or right.

---

* As we'll see later, these "bit inverters" are used for subtracting binary numbers.

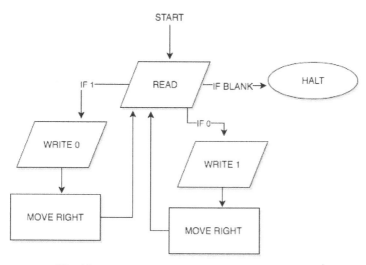

Fig. No. 2. Flowchart of Turing Program No. 1-2.

But the program still has only a single state, the initial state. Therefore, the Turing machine needed to run the instruction set is termed a one-state Turing machine. But more states permit more complicated operations, such as addition and subtraction of binary numbers.

Turing machines have also spawned their own brand of interesting programming puzzles. Here's a famous one, called the Busy Beaver Problem: How much data (0s or 1s) can a Turing machine write to a completely blank tape *and yet still* halt? Of course, without the halting restriction, a Turing machine could write data forever on a blank tape:

Turing Machine Program No. 1-3: A String of 1s Forever

STATE	READ	WRITE	NEXT MOVE	NEXT STATE
1	Blank	1	R	1
1	0	1	R	1
1	1	1	R	1

But our Busy Beaver needs to stop. Which means we can have, at most, one symbol printed on the tape for a one-state machine:

Turing Machine Program No. 1-4: The 1-State Busy Beaver

STATE	READ	WRITE	NEXT MOVE	NEXT STATE
1	Blank	1	R	Halt
1	0	1	R	Halt
1	1	1	R	Halt

What about a two-state machine? Take a look at the program shown on the top of the next page.

Turing Machine Program No. 1-5: The 2-State Busy Beaver

STATE	READ	WRITE	NEXT MOVE	NEXT STATE
1	Blank	1	R	2
1	0	Blank	None	1
1	1	1	L	2
2	Blank	1	L	1
2	0	Blank	None	1
2	1	1	R	Halt

Four 1s are written onto the paper tape; the machine takes six steps to get there before halting. Examine the flowchart for the program, which is quite a bit cluttered.*

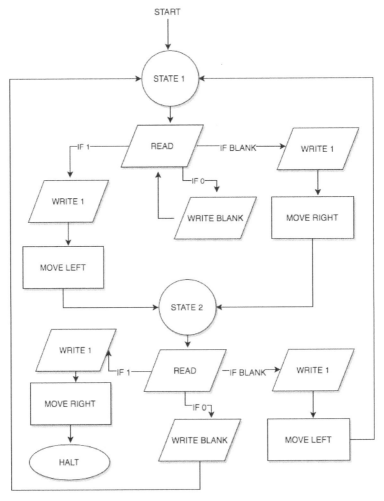

Fig. No. 3. Flowchart of Turing Program No. 1-5.

---

* Computer scientists compress these unwieldy flowcharts into what are called Turing machine state transition diagrams, simplifying following the steps of the instruction set quite a bit.

Here is, step by step, how the tape is processed by the read/write head:

B	B	B	B	B	B	B	B	B	B	B	B	B	B	B
				^										

B	B	B	B	1	B	B	B	B	B	B	B	B	B	B
			^											

B	B	B	1	B	B	B	B	B	B	B	B	B	B	B
			^											

B	B	B	1	1	B	B	B	B	B	B	B	B	B	B
			^											

B	B	B	B	1	1	B	B	B	B	B	B	B	B	B
			^											

B	B	B	B	B	1	1	B	B	B	B	B	B	B	B
			^											

B	B	B	B	1	1	1	B	B	B	B	B	B	B	B
			^											

B	B	B	B	B	1	1	1	B	B	B	B	B	B	B
			^											

B	B	B	B	1	1	1	1	B	B	B	B	B	B	B
			^											

B	B	B	1	1	1	1	B	B	B	B	B	B	B	B
			^											

With three states, the Busy Beaver instruction set looks like this:

Turing Machine Program No. 1-6: The 3-State Busy Beaver

STATE	READ	WRITE	NEXT MOVE	NEXT STATE
1	Blank	1	L	2
1	0	Blank	None	1
1	1	1	None	Halt
2	Blank	Blank	L	3
2	0	Blank	None	1
2	1	1	L	2
3	Blank	1	R	3
3	0	Blank	None	1
3	1	1	R	1

The 3-State Busy Beaver program will write six 1s in a row on the tape, but will do so circuitously, taking fourteen steps until halting. Note that there is no way to write an instruction set for a three-state machine that writes more than six 1s but still halts. A 4-State Busy Beaver program will print thirteen 1s, but it will take 107 steps to get there. From there, as the number of states increase, the number of steps prior to halting grows wildly.

paper computer design implications ◆◆◆◆◆◆◆◆◆◆◆◆◆◆◆◆◆◆◆◆◆◆◆

The CARDIAC, LMC, and Instructo are all finite-state machines, just like a Turing machine. But none of the three paper computers is Turing complete. Although they can simulate a single-tape Turing machine to some extent, there are physical limits that bog down all physical computer hardware, electronic or otherwise— limits of storage space and wear-and-tear— that the paper computers also succumb to. (Most computer languages, though, are in theory not subject to physical limits and therefore satisfy Turing completeness.)
◆◆◆

What is remarkable is that Alan Turing sketched out these ideas for a programmable digital (binary) computer, capable of computation, decades before such as a machine was actually constructed. But Turing did more than that. He also proposed the following: any instruction set that can be carried out by a Turing machine can also be carried out by a Universal Turing machine. In effect, a Universal Turing machine can simulate the operation of any other Turing machine simply by reading in the instruction set of the Turing machine. So a Universal Turing machine can be programmed to act as a whole host of Turing machines.

Now consider the set of all irrational numbers, numbers which cannot be written as fractions or decimals with any predictable pattern. Alan Turing proved that the Universal Turing machine can use a finite instruction set to compute *some* irrational numbers out to as many decimal places as desired, such as π or e (Euler's constant)— which are irrational but nonetheless have set algorithms for producing their digits to any nth decimal place. He called such numbers computable: "The 'computable' numbers may be described briefly as the real numbers whose expressions as a decimal are calculable by finite means," he wrote. But Turing also proved that there are many— in fact, infinitely many— numbers that are non-computable: no matter what finite instruction set is fed into his Universal Turing machine, correct decimal representations of this vast set of numbers cannot be generated. Turing connected this idea to the halting problem— Will a particular program stop on its own, or slip into an infinite loop and never halt?— and to Gödel, by noting that the halting problem is "undecidable," neither true nor false: there is no surefire finite algorithm that can reliably determine if a program halts or loops infinitely. The halting problem is a member of the set of non-computable not numbers but *functions*, which also break down into computable and non-computable. Since the halting problem proved undecidable (in a finite length of time— meaning, without potentially waiting forever to find out),[*] Alan Turing— and Alonzo Church, via other means— was able to answer Hilbert's *Entscheidungsproblem* in the negative, effectively halting Hilbert's Program.

In the years after the war, Turing continued to work on computing design, both in theory and in practice. He ran afoul of British law, however, and was

---

[*] The n-state Busy Beaver function is also non-computable.

convicted of "gross indecency," choosing a regimen of hormone treatments in lieu of imprisonment. In 1954, he was discovered dead in his apartment of apparent cyanide poisoning.

a short biography of john von neumann

Which brings us to the last name on our list of computer luminaries, John von Neumann. Born János Neumann in Hungary in 1903, he acquired the "von" when his father, a banker, was given the heredity title. Von Neumann demonstrated a strong faculty in mathematics and languages early on, but his father tried to steer him away from studying math for financial reasons. Nevertheless, von Neumann eventually earned a doctorate in mathematics from the University of Budapest.

Von Neumann would assume a number of university posts as lecturer, write papers about and with David Hilbert (right around the time Hilbert's Program was poised to collapse), and publish important works on quantum theory, game theory, group theory, set theory, and logic at a breakneck pace. His growing reputation quickly secured him a spot at the Institute for Advanced Study in Princeton. A decade after the appointment, von Neumann was invited to join the Manhattan Project, America's secret program tasked with constructing an atomic bomb during World War II.

By the end of the war, von Neumann turned his attention toward computers— specifically, the ENIAC, which he heavily modified, and a machine at the Institute for Advanced Study he designed, which was a binary arithmetic computer that stored program instructions and data in the same memory unit. Later on, he used cellular automata (mathematical shapes that are generated according to a strict set of discrete rules) to explore whether or not machines could potentially genetically reproduce. He died of cancer at age 53.

an (imaginary) interview with john von neumann

From who better to learn about the basics of von Neumann architecture than the man who did so much to lay the foundation for the computers we use today, John von Neumann? What follows is an imaginary interview I had with the man.

MARK JONES LORENZO: Thank you for sitting down with me today, Dr. von Neumann. Can I call you Johnny?

JOHN VON NEUMANN: No.

MJL: Okay. I was hoping you could describe some of the basics of the computer.

JVN: Very well. I will try to make it as simple as possible for you. You might think of a computer as having three components: a processor (or a CPU), a memory unit (or the primary memory or the main memory), and input/output devices.

MJL: Oh, you mean the peripherals, like a keyboard, monitor, printer, and mouse?

JVN: Why would you permit such an animal near the electronics?

MJL: I'm sorry, I forgot. You wouldn't know anything about a computer mouse, since you died in 1957, well in advance of its creation by Douglas Engelbart,[*] its use with GUIs at Xerox PARC, and its subsequent appropriation by Steve Jobs. And I'm not even sure you know what a monitor is— maybe you'd be familiar with the oscilloscope, which early computers used instead? Regardless, let's get back to the basic layout of the computer. The primary storage has another name, doesn't it?

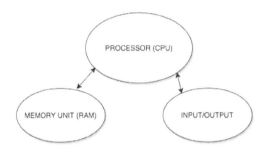

Fig. No. 4. Diagram of the basic layout of a computer.

JVN: Primary storage is commonly called RAM, standing for random access memory, which is volatile memory because it only holds data with a continuous power source. The switches or bits of RAM are arranged in groups of eight or sixteen or thirty-two. Of course, eight bits is called a byte. We store programs that the computer runs in RAM. Electrical signals transmitted from the CPU to RAM constitute a "write" operation. Signals which go in the other direction are called a "read" operation.

MJL: How are three components connected together?

JVN: Like in a large city, in which there needs to be some means of transportation from one place to another to keep goods and services and people freely and efficiently moving about, a computer has a system bus which intercon-

---

[*] Who famously gave a presentation in 1968 called "The Mother of All Demos," demonstrating the interplay between future computer hardware and software components like the mouse, windows, hypertext, video conferencing, and the like.

nects the three components. A bus is the wiring or electrical signals or connectors that transfer data between the three components, and supply these components with power.

MJL: And there are different types of busses?

JVN: Yes, you've made a surprisingly correct assumption. First, there is the address bus, which connects the processor (CPU) to the main memory or an input/output device. The wider the address bus, the greater the addressing capacity. Second, there is the control bus, which transmits control signals like reading, writing, and interrupting. Third, there is the data bus, which transfers data between the processor and the main memory or input/output device. The wider the data bus, the more data can be transferred at a time. Sometimes data busses only allow data to flow in one direction. For example, it would not make sense for a data bus to flow *from* an output device back to main memory. The address busses also send signals in only one direction.

MJL: But the main memory can store both data and program instructions.

JVN: Yes, in principle, there is no distinction between them. Unlike the Harvard model, in which instructions and data are separated.

MJL: Tell me about the processor.

JVN: The processor? Well, that is obviously where the arithmetic occurs, the operations on numbers. Again, we also call the processor a CPU.

MJL: What are the components of the processor?

JVN: There are the data registers, which store numbers, temporarily, that will be used for some sort of computation. There is the arithmetic and logic unit, or ALU, which performs the arithmetic on the data registers. There is the control unit, which controls or directs the operations of the ALU. There is a system clock, which generates pulses to help keep the control unit on task. Did I already mention the data registers?

MJL: Yes.

JVN: Okay, understood. There is also the instruction counter, which clues in the control unit on where it can find the next instruction to execute, and the instruction register, which is a copy of that "next instruction" that the control unit sees. And there is an interrupt register that signals when something else needs to be taken care of immediately by the CPU, like a button press from somewhere on a terminal. All of these components are involved in processing a single instruction.

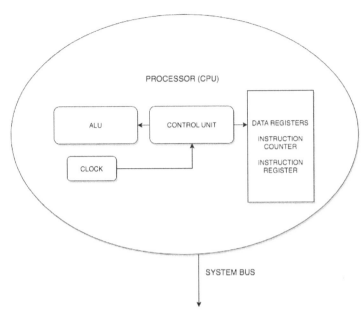

Fig. No. 5. Diagram of a computer processor.

MJL: So how does the instruction cycle work?

JVN: The instruction cycle, also called the execution cycle, has three steps, plus one more: fetch, decode, execute, plus check for interrupts. First, the instruction needs to be fetched from the main memory and placed into the instruction register. Next, the machine language instruction needs to decoded. Finally, after the decoding, the control unit sends a command to the ALU to perform the appropriate arithmetic or logical operations on the data in the registers, the results of which are stored in main memory. The entire execution cycle is trigged by a single pulse from the system clock. The higher the clock frequency, measured in Hertz, the faster the instruction execution time. Instructions are executed sequentially.

MJL: But it's not always the case that data and instructions are fetched from main memory, right? For example, there's high-speed cache memory.

JVN: I suppose in theory, although I wouldn't be inclined to take your word for it.

MJL: Okay, fine. And all of the arithmetic that the ALU does is in binary, right?

JVN: Yes. Let me explain why by reading some of my *First Draft of a Report on the EDVAC*,[*] explaining the design of what would later be called the von Neumann architecture. "Every digital computing device contains certain relay like elements, with discrete equilibria. Such an element has two or more distinct states in which it can exist indefinitely...."

MJL: But can't relay elements be mechanical, like moving gears?

JVN: I am about to get to that. Please let me continue reading. "In existing digital computing devices various mechanical or electrical devices have been used as elements: Wheels, which can be locked into any one of ten (or more) significant positions, and which on moving from one position to another transmit electric pulses that may cause other similar wheels to move; single or combined telegraph relays, actuated by an electromagnet and opening or closing electric circuits; combinations of these two elements; and finally there exists the plausible and tempting possibility of using vacuum tubes, the grid acting as a valve for the cathode-plate circuit. In the last mentioned case the grid may also be replaced by deflecting organs, i.e. the vacuum tube by a cathode ray tube— but it is likely that for some time to come the greater availability and various electrical advantages of the vacuum tubes proper will keep the first procedure in the foreground.... It is clear that a very high speed computing device should ideally have vacuum tube elements."

MJL: So that's why vacuum tubes became dominant: their speed.

JVN: Precisely. But I still have not answered your initial question, about the use of binary arithmetic. Continuing: "Thus, whether the tubes are used as gates or as triggers, the all-or-none, two equilibrium, arrangements are the simplest ones. Since these tube arrangements are to handle numbers by means of their digits, it is natural to use a system of arithmetic in which the digits are also two valued. This suggests the use of the binary system."

MJL: "Suggests the use...." But was actually *using* binary simpler?

---

[*] http://www.wiley.com/legacy/wileychi/wang_archi/supp/appendix_a.pdf

JVN: It most definitely was, in terms of arithmetic calculations. Now please stop interrupting as I read one more passage from my report.

MJL: Go ahead.

JVN: "A consistent use of the binary system is also likely to simplify the operations of multiplication and division considerably. Specifically it does away with the decimal multiplication table, or with the alternative double procedure of building up the multiples of each multiplier or quotient digit by additions first, and then combining these (according to positional value) by a second sequence of additions or subtractions. In other words: Binary arithmetics has a simpler and more one-piece logical structure than any other, particularly than the decimal one." Now you may comment.

MJL: But binary doesn't come naturally to human beings; decimal does.

JVN: A small price to pay for the increase in speed. Binary is the simplest possible number system. In addition, binary calculations sped up considerably with the later invention of the transistor, a natural fit for the "on" and "off" states of binary logic.

MJL: Of course, with your architecture there's also, what's called now, the von Neumann bottleneck— a limitation caused by the computer having to fetch instructions before actually running them.

JVN: Why do you have to bring that up? You are terribly frustrating to speak with. I should have never agreed to this interview, the contents of which you will probably use to write some Whiggish history of computing.

MJL: I apologize; I didn't mean to offend you. Look, so there's obviously more complexity than what you've covered with me here. Computers aren't restricted to main memory, and can read and write to a secondary storage device, for instance.

JVN: In theory, yes. Describing all of that would add layers of complexity to our discussion. But I instead wanted to give you a simple overview of how and why computers, as we know them today, came about, which you barely let me do.

MJL: But you did do it, and you did it well. Now I'll be better able to understand the decisions that the creators of the CARDIAC, LMC, and Instructo made when modeling computer architecture, since all three paper computers are von Neumann architecture stored-program computers.

JVN: What is an "Instructo"?

MJL: Never mind. Thank you for your time.

JVN: Humph.

MJL: One more thing before you go. Can I have your autograph? Or, better yet, your brain?

paper computer design implications ♦♦♦♦♦♦♦♦♦♦♦♦♦♦♦♦♦♦♦♦♦♦♦♦♦♦♦
The CARDIAC, LMC, and Instructo are all modeled after the von Neumann architecture of stored-program computers, rather than the Harvard architecture. Busses are only implied in the CARDIAC and LMC, but they are explicitly laid out on the front face of the Instructo.
♦♦♦

an introduction to binary

Real-world applications of the binary number system lie all around us, such as in UPC barcodes. But since most human beings don't have only two fingers, counting in binary doesn't come particularly naturally to us. Although all three paper computers dispense with binary (base 2) operations in favor of decimal (base 10) for simplicity's sake, let's nevertheless spend some time examining how computers work with binary so we will have a firmer grasp on computing basics.

Decimal wasn't humankind's first foray into number systems. The Egyptians, Babylonian,* and, perhaps the best known, Romans all created symbols for tabulating quantities. But when the decimal Hindu-Arabic system arrived, it made curiosities out of all number systems predating it— except for Roman numerals affixed onto fancy clock faces and past Super Bowls— because of three characteristics, as *Code: The Hidden Language of Computer Hardware and Software* author Charles Petzold explains: (1) The introduction of positional digits, meaning the location of the digit, in addition to the quantity the digit expresses, is meaningful; (2) Unlike other number systems, there is no designated symbol for ten; and (3) The presence of the zero digit, which functions as a placeholder.

Carrying digits is integral when working with a positional number system, whether in decimal or binary. When we find the sum of 1 and 1 in decimal, we don't need to carry. But when we find the sum of 1 and 9, we do: the ones' place flips to 0, and the tens' place flips to 1. This is called carrying— here, we carried a 1.

Whereas with decimal, we can count up ten consecutive digits before having to use two of them together— 0, 1, 2, 3, 4, 5, 6, 7, 8, 9— in binary, which is the simplest possible number system, we can only count up two: 0 and 1. Counting from 0 to 9 with binary gives us 0, 1, 10, 11, 100, 101, 110, 111, 1000,

---

* Base 60, or sexagesimal, but lacking the number zero.

1001.* Thus, binary arithmetic will require carrying with even the smallest decimal addition problems.

The table below summarizes both the "sum bit" (the direct result of the addition of two bits) and the "carry bit" (the bit that's carried over to the next column):

SUM BIT			CARRY BIT		
Inputs		Output	Inputs		Output
0	0	0	0	0	0
0	1	1	0	1	0
1	0	1	1	0	0
1	1	0	1	1	1

Because there are actually two bits of information— the sum bit and the carry bit— with every addition operation, the result of each operation can be found using this two-bit table:

Inputs		Output
0	0	00
0	1	01
1	0	01
1	1	10

For instance, let's add 6 and 7 in binary; see if you can spot the carries:

$$\begin{array}{r} 110 \\ +111 \\ \hline 1101 \end{array}$$

---

* Of course, there are more bases than just base 2 and base 10. For instance, there is base 8, or octal; counting with octal looks like this: 0, 1, 2, 3, 4, 5, 6, 7, 10 (because we've run out of unique symbols), 11, 12, 13, 14, 15, 16, 17, 20, ...

Note that converting 1101 to base 10 gives us 13, which is the sum of 6 and 7. Converting the binary number 1101 to the decimal number 13 is accomplished by using powers of two, one for each placeholder:

$$= 1 \cdot 2^3 + 1 \cdot 2^2 + 0 \cdot 2^1 + 1 \cdot 2^0$$

$$= 8 + 4 + 0 + 1$$

$$= 13$$

At this point, you may have realized that any decimal power of two automatically results in a binary number of 1 followed by zeros. That's because when converting the binary digits to decimal, only the highest power of two is multiplied by 1, while all the descending powers of two are multiplied by zero.

Converting from base 2 to base 10, however, requires a lot more work, since by using binary we're only counting by twos before shifting a place value to the left. We will have to repeatedly divide and take remainders, which can only be 0 or 1, into account. These remainders will constitute the binary digits of the converted decimal number.

Consider the number 84, for example. To convert 84 to base 2, we do the following:

$$2\overline{)84}, R = 0 \quad \text{quotient } 42$$

$$2\overline{)42}, R = 0 \quad \text{quotient } 21$$

$$2\overline{)21}, R = 1 \quad \text{quotient } 10$$

$$2\overline{)10}, R = 0 \quad \text{quotient } 5$$

$$2\overline{)5}, R = 1 \quad \text{quotient } 2$$

$$2\overline{)2}, R = 0 \quad \text{quotient } 1$$

$$2\overline{)1}, R = 1 \quad \text{quotient } 0$$

Writing the remainders in reverse order gives us the result: in base 2, 84 is **1010100**.

Modern computers hold binary numbers in registers. Registers, therefore, are storage compartments for addresses, instructions, or simply data— all in binary form. The size of the registers are measured in bits, short for binary digits.* A 1-bit register can hold either 0 or 1. A 2-bit register can hold twice that information: 00, 01, 10, or 11. A 3-bit register can hold twice that: 000, 001, 011, 101, 010, 100, 110, or 111. In general, then, an n-bit register can hold 2^n binary numbers.

Registers of the past typically had 8-bit storage. Eight bits is usually called a "byte," a term originating in the sterile business workspaces of IBM in the mid-fifties; byte was originally spelled "bite" but the y replaced the i to avoid confusion with the word "bit."[†] A byte can hold $2^8 = 256$ unique pieces of information: that's all binary numbers from 00000000 to 11111111, corresponding to the decimal numbers from 0 to 255. Bytes can also easily be converted to hexadecimal (usually abbreviated hex), which is base 16: 0, 1, 2, 3, 4, 5, 6, 7, 8, 9, A, B, C, D, E, F, 10, 11, 12, etc. The rightmost digit of a hexadecimal number counts in ones, like in decimal; but the digit immediately to the left in hex counts in 16s (decimal: in tens), to the left of that in 256s (decimal: in hundreds), to the left of that in 4096s (decimal: in thousands), and so on.

What's the point of hex? Although computers don't "think" in hexadecimal, converting between binary and hex, by grouping a byte into four bits at a time, is convenient to us humans who actually do think (usually in decimal, but hex isn't particularly counterintuitive). Each hexadecimal digit corresponds to *exactly* four bits— and all combinations of these four bits exhaustively match to each possible hex digit. The following table helps to make this clear:

Binary	Hex	Binary	Hex	Binary	Hex	Binary	Hex
0000	0	0100	4	1000	8	1100	C
0001	1	0101	5	1001	9	1101	D
0010	2	0110	6	1010	A	1110	E
0011	3	0111	7	1011	B	1111	F

For instance, if we wish to convert the binary number 01101111 into hex, we first split the eight-digit number into two groups: 0110 (the high-order byte) and 1111 (the low-order byte). Then we use the table to quickly match the

---

* Perhaps the statistician John W. Tukey coined the term— he was fond of playing with language for jargony, technical terms, such as the "side-foot," his neologism for 12-inch stacks of computer printout— but Claude Shannon, who attributes it to Tukey, was the first to use "bit"— after dispensing with "bigit" and "binit" as abbreviations a bit too long— in a published work in 1948. Either way, both Tukey and Shannon were working at Bell Labs at the time.

[†] A half-byte is called— what else?— a "nibble" or "nybble."

four-digit binary numbers with their corresponding hexadecimal counterparts: 6 and F, or 6Fh. The *h* appended to the value, of course, means hexadecimal.

Today's computers usually have 32-bit registers (thirty-two bits is called a "full word"), which has space for $2^{32} = 4,294,967,296$ unique pieces of information,* but to take into account negative values the four-billion-number spread has to be chopped in half: roughly two billion numbers set below zero, and two billion above it. These are called "signed binary numbers," because they account for positive and negative values.

But why both negative and positive values, rather than just positive? The idea is to avoid using the negative sign with binary numbers, and express all decimal numbers— positive and negative— using positive (sign-less) binary numbers. This means (1) Restricting the number of possible decimal numbers (as we did in the paragraph above— to 4,294,967,296 of them), and (2) Assigning roughly half of the binary digits to correspond with negative decimal numbers, using a process known as "two's complement."† Essentially, it all boils down to this: any negative decimal number is represented by a binary number with a leading digit of 1, while any positive (or zero) decimal number is represented by a binary number with a leading digit of 0. The leading digit of these binary numbers is called the "sign bit": 1 for negative, and 0 for positive or zero. Note that binary numbers can be "unsigned" as well, dispensing with the sign bit entirely in favor of space for nearly twice as many positive-only decimal number correspondences.

paper computer design implications ◆◆◆◆◆◆◆◆◆◆◆◆◆◆◆◆◆◆◆◆◆◆◆◆◆◆◆
The loose idea of the sign bit carries over with the CARDIAC and LMC, specifically with respect to their accumulator sign tests for conditional jumps.
◆◆◆

By the way, it's easy to work backwards, too: you can calculate how many bits of storage are required for a set of data by taking the base two logarithm of the quantity. For example, to find out how much storage is necessary for 8,192 pieces of information, simply find $\log_2 8192$, which is 13. Thus, 13 bits. Even if we wanted to represent only, say, 8,000 pieces of information, we'd still need 13 bits, since $2^{12} = 4,096$, which is less than 8,000.

Computers process binary arithmetic, such as addition, by holding one number in one register, another number in a second register, and computing the sum in yet a third register. More complex arithmetic can also be per-

---

* The number of unique pieces of information is usually called the "resolution."

† To find the two's complement, invert the bits of the binary number, and then add 1. For example, consider 010. First, flip the bits: 101. Then, add 1: 110.

formed analogously. (The ENIAC utilized octal-base radio tubes, but performed arithmetic with decimal-based accumulators.)

paper computer design implications ♦♦♦♦♦♦♦♦♦♦♦♦♦♦♦♦♦♦♦♦♦♦♦♦
The CARDIAC and LMC dispense with binary in favor of decimal for simplicity's sake. Thus, hexadecimal conversions are rendered moot. The Instructo, though mostly running on decimal as well, does have a bit of binary logic "wired in" to its three Jump Switches (three bits' worth, anyway). You might think of the value of each Jump Switch— 0 (false) or 1 (true)— as a one-bit Boolean "flag."
♦♦

george boole and his boolean algebra

The mathematician George Boole was born in England in 1815, the same year as Ada Lovelace. With a little help from his father, he taught himself classical languages and mathematics and was ultimately appointed a professor of mathematics in 1849 in Ireland. Boole's interest was in mathematical logic; he published a number of books on the subject before succumbing to pneumonia at age 49.

The ancient Greeks had fashioned deductive logic as a gateway to the truth, but their vehicle had been the study of language, courtesy of the syllogism. Their most famous syllogism— in which a conclusion follows two explicitly stated premises— is *All men are mortal. Socrates is a man. Therefore, Socrates is mortal.* But Boole instead felt that mathematics, not language, was the Rosetta stone for systematizing logic; he even believed that this mathematical systematization could shed light on the inner workings of the brain.

Boole wrapped logic in the symbols of algebra: the operands (the numbers or variables, like x and y) and the operators (the symbols that represent mathematical operations, such as + and ×). Algebra also has properties that hold when manipulating numbers and symbols, such as commutativity (the order of operations don't affect the result, which is true for addition and multiplication), associativity (grouping symbols don't affect the result, which is also true for addition and multiplication), and the distributive property.

Boole abstracted algebra, leaving the basic rules of commutativity, associativity, and the distributive property intact, but he designated the operands to refer to classes or sets of things rather than variables (like x or y) or numbers.

In the context of sets, though, the + and × signs had to be repurposed. The addition sign now referred to a union of two sets, meaning all elements of both sets combined. For example, if set A is of all red Volkswagen Beetles and set B is of all white Volkswagen Beetles, then the set $A + B$ includes Volkswagen Beetles that are *either* red or white (or both, of which a "two tone" Beetle technically is possible).

The multiplication sign refers to an intersection of two sets, meaning only the elements in common between the two sets. Consider set B, all red

Volkswagen Beetles, and set *C*, all Volkswagens. The set *B* × *C* contains only the common elements: the red Volkswagen Beetles.*

In addition to unions and intersections, we need to understand the symbol 1, which refers to the entire set of objects in context. Suppose that set *D* contains all other cars besides Volkswagens. Then, *C* + *D* = 1, the set of all cars. It is also the case that 1 − *C* = *D*, meaning that the complement of set *C*, all Volkswagens, is equivalent to set *D*, all other cars besides Volkswagens.

The symbol 0 represents the empty, or null, set— which is the set containing nothing at all. If we wanted to find the intersection of *C* and *D*, we would come up empty-handed because there is nothing in common between them. Thus, *C* × *D* = 0.

Boole realized that he could use these mathematical rules of logic— along with an already well-established rule, Aristotle's law of excluded middle: "it will not be possible to be and not to be the same thing," i.e., either *A* or not *A* is true— to translate linguistic syllogisms like *All men are mortal...* into mathematical statements using the operands and operators of sets. By doing so, we can confirm that Socrates was mortal, thankfully.

But we can also bring language back into the fray, as Charles Petzold concisely elaborates in his book *Code*. Instead of + to represent union, use the word OR— since, after all, a union enumerates all elements from either the first set *or* the second set.† And, instead of using ×, use the word AND— since an intersection represents all that is in common between the first set *and* the second set. Finally, when finding the complement (by using the symbols 1 −), you are really capturing everything *not* in the set— thus, use the word NOT.

We might consider the following real-world examples to help us better understand OR, AND, and NOT. If you are unfortunately involved in a minor car accident, when two uniformed police officers show up at the scene they will pair off: one officer querying you about the details, while the other asks the opposing party. Both officers will ask their respective parties: Do you wish to file a police report? If either you OR the other party (or both) answers yes, then a report is filed.

The AND operator comes into play at a wedding ceremony: only if both the bride AND the groom say "I do" is the couple officially declared to be married.

Finally, the NOT operator pops up any time a complement is required. For example, if I ask a room full of people, "Who is NOT left-handed?" then I am asking for more than just right-handed people to identify themselves; rather,

---

* Note that the symbols ∪ and ∩ can also be used to represent union and intersection, respectively.

† More specifically, an *inclusive or*, since the "or" accounts implicitly for all elements in either *or* both sets.

anyone who is right-handed, ambidextrous, no-handed, foot-handed, head-handed, etc., must be called to account.

If we pare Boolean algebra down to just two operands, 0 (representing "False" or "Off") and 1 (representing "True" or "On"), then we can make a key conceptual leap: the switches, relays, circuits, and the like powering computers conform to the logic and operators of Boolean algebra. Petzold laments that this conceptual leap took much longer than necessary. "But nobody in the nineteenth century made the connection between the ANDs and ORs of Boolean algebra and the wiring of simple switches in series [physically representing ANDs] and in parallel [physically representing ORs]," he writes. "No mathematician, no electrician, no telegraph operator," not even Charles Babbage, who, despite corresponding with Boole, never realized that he would have been better served by ditching much of the physical implements of his computers in favor of the on-off binary function of telegraph relays.

augustus de morgan and his eponymous laws

Before we explore how to use OR, AND, and NOT with binary digits, we should take a moment to mention Augustus De Morgan's contributions to Boolean algebra. Born in India in 1806, De Morgan grew up in England and attended some of its finest schools. He went on to become a professor of mathematics at University College in London, and arrived at a great many groundbreaking results, including the formalization of mathematical induction and advances in complex number theory. De Morgan also contributed to the study of abstract algebra and logic, but he is best known for his eponymous laws. The first law is

$$(1-X) \times (1-Y) = 1 - (X+Y)$$

Using a Venn diagram, the first law looks like this:

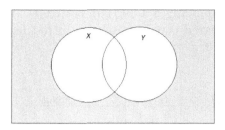

The second law is

$$(1-X)+(1-B)=1-(X\times Y)$$

Using a Venn diagram, here is the second law:

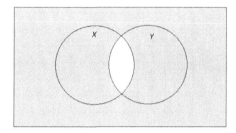

Though De Morgan's contributions were great, the elegant simplicity of his laws makes them his most memorable mathematical result.

some bitwise operations

Reconsider bits, which can take one of two values: 0 or 1. Besides with traditional arithmetic (as we saw several pages back), bits can be manipulated, bit by bit, using bitwise operations— and bitwise operations are faster for a computer to perform than binary arithmetic. Some of the following bitwise operators (along with their truth tables) should now be familiar to you; others, though, we haven't yet encountered (and they postdate George Boole).

First, there is bitwise NOT operator, which inverts each bit: e.g., **1010100** (decimal: 84) becomes **0101011** (decimal: 43). Inverting bits in this way is also characterized as finding the complement.

NOT	
Input	Output
0	1
1	0

Next, the bitwise AND operator compares pairs of bits in the same position— e.g., the first digit of two binary numbers, or the fourth digit of two binary numbers— outputting a **1** *only* if both bits are **1** or a **0** if either (or both) bits being compared is **0**. For instance, **010** (decimal: 2) AND **111** (decimal: 7) is equal to **010**.*

AND	
Inputs	Output
0 0	0
0 1	0
1 0	0
1 1	1

Similarly, the bitwise OR operator compares bits in the same position, but outputs a **1** if *either* bit is a **1**, or a **0** otherwise. For example, **010** OR **111** is equal to **111**.

OR	
Inputs	Output
0 0	0
0 1	1
1 0	1
1 1	1

We could *negate* both AND and OR by taking the complement of both, resulting in the NAND and NOR operators.

NAND			NOR	
Inputs	Output		Inputs	Output
0 0	1		0 0	1
0 1	1		0 1	0
1 0	1		1 0	0
1 1	0		1 1	0

Finally, there is the bitwise XOR operator, which checks to see if bits are the *same* or *different*. (It's formally called an "Exclusive OR" because the output of 1 is restricted to *either* compared bit equaling 1, but *not both of them equaling 1*.) So, **0101** XOR **1100** results in **1001**, since only the middle digits match. The XOR table is shown on the next page.

---

* It's like that old joke: 7 and 2 walk into a bar, and out walks 2.

XOR		
Inputs		Output
0	0	0
0	1	1
1	0	1
1	1	0

Recall the adding of binary digits: each addition operation requires two bits, a sum bit and a carry bit. If you reexamine the carry bit table from several pages ago, you'll see it is the AND operator in disguise. And if you look again at the sum bit table, you will find that it's simply the XOR operator. With a little ingenuity and whole lot of wiring, you can construct a set of full adders (made of half adders, which are themselves comprised of logic gates and relays)[*] that electronically add together binary numbers by using the Boolean principles laid out above.

Binary subtraction can be performed by using a complement, courtesy of a bit inverter, and then adding, thus completely avoiding any borrowing.

paper computer design implications

Unfortunately, neither the CARDIAC nor the LMC allows for any of these bitwise operations, although there are ways to make simple comparisons. The Instructo, courtesy of its Compare Unit, permits the direct comparison (greater than, less than, or equal to) of values stored in its registers, but it also does not permit Boolean comparisons— despite its binary Jump Switches.

the supreme court votes bitwise

Imagine the U.S. Supreme Court as a 9-bit register, one bit for each justice. If the Supreme Court votes 5-4 to strike down some issue— call it Statute A— then the NOT (Vote on Statute B) would reverse the decision, since votes in favor would turn against and vice versa.

Now suppose there are two Statutes in question: A and B. The voting for Statute A was YNYYNYYNN and for Statute B was NYNNNYNYN. Let's use bitwise operators to package the votes together:

(YNYYNYYNN) AND (NYNNNYNYN) → (NNNNNYNNN)

(YNYYNYYNN) OR (NYNNNYNYN) → (YYYYNYYYN)

(YNYYNYYNN) XOR (NYNNNYNYN) → (NNNNYYNNY)

---

[*] As Ron White, in his book *How Computers Work*, notes, "More than 260 transistors are needed to create a full adder that can handle mathematical operations for 16-bit numbers."

So, if both Statutes are considered together, the AND comparison results in a 8-1 decision striking down them down, the OR comparison results in a 7-2 decision upholding them, and the XOR comparison results in a 6-3 decision striking them down.*

a bitwise operators game

Instead of the Supreme Court, imagine a society in which golden coins, each of the same denomination, only count as currency if they land on the ground in the heads position. To pay for any item, you have to throw eight golden coins on the ground at a time. When the coins land, they are arranged by placing all of the coins that land on heads next to each other, followed by all the coins that land on tails.

If not enough coins land on heads to pay for your item, you are allowed to use the bitwise operators: NOT, if the complement will help you,† or AND, OR, XOR— or a linear combination thereof— but you'll have to throw a second stack of eight coins on the ground to compare with your first stack.

Suppose an item costs 7 golden coins. You throw your stack and are short two coins: HHHHHHTT. The NOT operator won't help here, because then you'll be short four coins. You have no choice but to throw a second stack of eight coins. What are the possible combinations of heads and tails for the second stack that allow you sufficient funds to purchase the item if: (1) You are only allowed to use AND, OR, or XOR once; or (2) You are free to use any combination of AND, OR, XOR, and NOT, in any order, any number of times? Of course, the second question will have numerous answers. Another point to consider: If we change the cost of the item, or the number of coins in a thrown stack, how does the answer or answers change? Finding the solutions to these questions is left as a challenge to the reader.

bit shifts

There are bitwise operators called bit shifts, which literally "shift" the place values of the binary number in a register, either left or right, substituting in 0s or 1s for the "missing" digits since a register must keep constant the number of digits. Consider an eight-digit binary number lodged in an 8-bit register, like **01100011**. The result of a left arithmetic shift (by one digit) is **11000110**, while a right arithmetic shift (also by one digit) would instead result in **10110001**. Notice that the left shift occasioned a fresh 0 at the tail end of the binary number, while a 1 was appended to the front of the binary number after the right shift. The leading 1 keeps the sign of the number intact post-right-shift.

---

* Realize that we're assigning a Y vote to represent 1, and an N vote represent 0.

† You would flip all eight coins over, e.g., HHTTTTTT would turn into HHHHHHTT.

Shifts aren't restricted to pushing only one digit out of the way— they can left or right shift multiple digits at a time. In addition, unlike arithmetic shifts, logical shifts insert only 0s to replace shifted digits, meaning that signs of numbers are lost after a right logical shift.

Logical bit shifting to the right is the equivalent of dividing by powers of 2. Take **10000000** (decimal: 128). A logical right shift (by one digit) results in **01000000** (decimal: 64). Notice that 128 divided by two is 64.

paper computer design implications ◆◆◆◆◆◆◆◆◆◆◆◆◆◆◆◆◆◆◆◆◆◆◆
The CARDIAC, but not the LMC or Instructo, has an opcode that allows it to automatically shift a value lodged in the accumulator by a set number of digits— but in decimal, not binary, and the missing digits are always replaced by zeros. Shifting on the LMC or Instructo necessitates writing a program.
◆◆

All of these binary operations— addition, subtraction, multiplication, division, NOT, AND, OR, XOR, and bit shifts— are the responsibility of the arithmetic and logic unit (ALU), which performs operations on the data registers. But each time an operation is set to take place, the ALU has to fetch data from memory and copy it to registers, and store arithmetic results into yet another register. Which is fine for finding the sum of two numbers, but what about calculating the sum of ten numbers? Recall the central conceit of the Little Man Computer: there exists a little man whose job it is to race back and forth inside the computer, reading inputted data into addresses, writing results to output, and otherwise doing all the grunt work necessary to keeping the computer operational. But if the little man has to run from the ALU back and forth to multiple registers for a simple addition problem, then he'll quickly tire himself out (little men not having the stamina of us full-size people).

So what if, instead of placing binary numbers in multiple registers— which have to be fetched from memory first, operated on, and then, after the answer is found, copied into another register— all the calculations take place inside just *one* dedicated register, called an accumulator, which stores the intermediate steps of the ALU? Our little man can save his energy since he won't have to race around nearly as much.

paper computer design implications ◆◆◆◆◆◆◆◆◆◆◆◆◆◆◆◆◆◆◆◆◆◆◆
It makes sense for the Little Man Computer to make use of a single accumulator, instead of multiple registers, for simplicity's sake (and the little man's long-term health).
◆◆

an accumulator running the show

Accumulators are, in essence, general-purpose arithmetic registers and were present in the very first electromechanical computers. For example, in addition to retaining 60 registers for the input of numerical data, the Harvard Mark I had an additional 72 registers which could both store numbers (up to 23 digits, along with their sign) and perform simple arithmetical operations (addition and subtraction) on them. The ENIAC also utilized accumulators to perform calculations.

In a modern computer with an accumulator-based instruction set architecture,[*] as opposed to a general-purpose register (GPR) instruction set architecture, the opcodes (machine language commands) are much simpler to write since numerical values are loaded, stored, and added in a single place: the accumulator. Thus, opcodes for accumulator-based instruction sets require only a single operand, which is the object or data that will be operated on; the accumulator never needs to be specified in the opcodes' arguments. As the text *Write Great Code, Vol. 2: Thinking Low-Level, Writing High-Level* by Randall Hyde makes clear,

> The idea behind an accumulator-based machine is that you provide a single accumulator register, where the CPU compares temporary results, rather than computing temporary values in memory.... Accumulator-based machines are also known as *one-address* or *single-address machines* because most instructions that operate on two operands use the accumulator as the default destination operand and require a single memory or constant operand to use as the source operand for the computation. [author's italics]

By contrast, most opcodes for a general-purpose register machine necessitates multiple arguments, loading values into various registers and performing operations on them— and storing those operations in yet more registers. (To be fair, general purpose registers can double as accumulators themselves, saving memory.) Muddying the waters even further, there are different types of GPR architectures, such as register-register,[†] register-memory, and memory-memory, which are categorized by the functionality of the operands. In spite of the additional memory demands, there are a variety of advantages of GPR architectures, such as flexibility and— since values for multistep calculations don't need to be repeatedly reloaded into an accumulator— ease of use for more complex calculations, the little man's health notwithstanding.

paper computer design implications ◆◆◆◆◆◆◆◆◆◆◆◆◆◆◆◆◆◆◆◆◆◆◆◆◆◆◆◆

Unsurprisingly, to keep things as simple as possible, the CARDIAC and the LMC are both single-accumulator, single-address machines. The Instructo, on the other hand, has multiple general-purpose registers and thus is a GPR machine— necessitating a more complex instruction set than either the CARDIAC or LMC. Nonetheless, the

---

[*] Such as the 6502 microprocessor.

[†] Such as the ubiquitous x86 microprocessor.

accumulators of the CARDIAC and LMC can only handle relatively small (decimal) numbers. Specifically, the CARDIAC/LMC accumulator can handle four digits, despite the contents of memory cells permitting at most three digits, for the express purpose of overflow (i.e., when the calculated sum of two numbers exceeds 999). The Instructo, on the other hand, has no explicit limitation on the size of numbers stored in its registers.

❖❖❖

how a computer remembers—and forgets

Modern computers operate according to a complex hierarchy of memory storage. Storage can be permanent/persistent (ROM, or read-only memory) or temporary (RAM, or random access memory; addresses in memory that are "random access" don't need to be read sequentially— i.e., from the first address to the last).

The idea behind ROM is simple: if there is to be such a thing as "permanent" memory, then even if power is lost to the device, data needs to remain intact; microchips with this property are called nonvolatile. In addition, the data stored on ROM chips needs to be unchangeable— or nearly so. For instance, when I turn on my old Tandy Color Computer (CoCo) 2, within less than a second it boots straight to Microsoft BASIC— because a BASIC interpreter is encoded directly onto ROM. These ROM chips, like those in greeting cards that play music when they're opened, are unchangeable.

There are various kinds of ROM, some less immutable than others. Consider PROM (programmable read-only memory), which can be programmed once; EPROM (erasable programmable read-only memory), which can be rewritten repeatedly using an ultraviolet (UV) light; and EEPROM (electrically erasable programmable read-only memory), which dispenses with the UV light in favor of an electric field for rewriting. Flash memory is a type of EEPROM, although it's modified to speed up the rewriting process.[*] Note that paper tape and punched cards were also examples of persistent memory, as were magnetic storage devices like floppy disks and cassette tapes (the CoCo 2 used a run-of-the-mill cassette player for read/write memory storage), since data is retained without a continuous power supply.

Conversely, RAM is volatile memory, because power is required to keep the contents of the data intact. In the past, magnetic cores, made of ferromagnetic and ceramic material, were utilized, as were tubes of mercury. Computers today can make use of DRAM (dynamic random access memory). DRAM contains many memory cells. Each of these memory cells has an address, located by taking the intersection of the column (called a bitline) and the row (called a wordline) on an etched silicon wafer chip. Each memory cell contains a transistor and a capacitor that can be set to 0 by emptying a "bucket" of electrons, or to 1 by filling a "bucket" of electrons. But the contents of DRAM

---

[*] http://computer.howstuffworks.com/rom.htm/printable

need be to constantly maintained by a memory controller while the computer is operational to avoid errors, such as memory leaks, which occur if the capacitors in the memory cells aren't repeatedly refreshed before discharging to 0 by emptying their electron "buckets," as they are wont to naturally do.[*]

There are also two memory registers we will briefly mention. The MAR (Memory Address Register) stores the location (i.e., the address) of an instruction or of data. The MDR (Memory Data Register) holds data that will be written to or read from memory.

paper computer design implications ◆◆◆◆◆◆◆◆◆◆◆◆◆◆◆◆◆◆◆◆◆◆◆◆◆

Neither the LMC nor the Instructo makes explicit use of ROM, but both have mechanisms to fetch instructions and write to RAM with one hundred memory cells accessible by two-digit addresses (that do not need to be accessed sequentially, from the first address to the last). The CARDIAC, however, has ninety-eight memory cells of RAM accessible by two-digit addresses as well as two memory cells dedicated to ROM: memory cell 00 (read first by the CARDIAC, since the machine reads addresses sequentially, in ascending order, by default), which contains the word "001" hardwired into permanent memory, and memory cell 99, which contains the word "8--" in permanent memory. Since cell 99's operand, but not opcode, can be modified, it perhaps can be loosely thought of as EPROM or EEPROM. In addition, the CARDIAC and LMC make explicit use of a MAR, with space for two-digit addresses, and an MDR, with space for three digits, while the Instructo implicitly uses a MAR and an MDR but without the digits' size restrictions.

◆◆◆

pulling oneself up by one's bootstraps

What we call the beginning is often the end
And to make an end is to make a beginning.
The end is where we start from.
           — T. S. Eliot, "Little Gidding"

When you turn your computer on, it boots up by running a bootloader program. The bootloader is assigned the task of performing power-on self-tests (POST) and loading the operating system and other system software into computer memory.

Software is loaded by other software, but this results in an infinite regress, where the beginning is the end and the end the beginning: there has to be some way of loading that *initial* piece of software into the computer. A computer must lift itself up by its own bootstraps; it must find a way to bootstrap, shortened to boot, or self-start. Read-only memory offers the solution by containing nonvolatile and unchangeable start-up programs, but early computers didn't have that advantage.

---

[*] http://computer.howstuffworks.com/ram.htm/printable

For example, a bootloader might be loaded into main memory (RAM) from permanent or persistent memory (ROM) by means of a hard disk. But other types of persistent memory include magnetic tape or punched cards, both of which can contain bootloaders. Early IBM computers called the booting process an Initial Program Load (IPL), and started up by loading the contents of punched cards through input.

However, note that when a computer is first turned on, it's not necessarily the case that there's "nothing" inside each memory cell; rather, memory cells and registers are set, by mere chance alone, to arbitrary zeros or ones and then, later, their contents are (perhaps) overwritten by a running program.

paper computer design implications ♦♦♦♦♦♦♦♦♦♦♦♦♦♦♦♦♦♦♦♦♦♦♦♦♦♦

Thanks to the word "001" stored in its ROM, the CARDIAC permits a bootstrapping routine, although— this being not a real but a paper computer— the process of program loading can, and usually will, be shortcut. The LMC and Instructo both require a great deal of jury-rigging in order to build functional bootloaders, since all one hundred memory cells in each are RAM. The Instructo does, though, have a designated Start/Stop Switch as well as a Reset/Clear Switch (which clears the registers and resets the counters) to at least get you thinking about the boot-loading process.

♦♦♦

coding, high and low

Code offers us a way to communicate between us and our computers, whether it be by English words and phrases— à la COBOL (COmmon Business Oriented Language, developed in part by Grace Hopper), FORTRAN (FORmula TRANslation), ALGOL (ALGOrithmic Language), BASIC (Beginner's All-purpose Symbolic Instruction Code), Ada (named after Ada Lovelace, of course), or some other high-level computer language— or the 0s and 1s of machine language. Regardless, the words and syntax of any computer language must be concrete and unambiguous: the same term cannot have multiple meanings that change based on context, unless those changes (and contexts) are explicitly defined in advance. A computer, though, doesn't understand any programming language directly; rather, a compiler, which is a special kind of computer program, is needed to translate the programs we write in a high-level algorithmic language into a machine language that can be read by the computer's hardware.[*]

But if we strip things down to a bare minimum, what codes do we need to communicate with a computer? It depends on what we mean by "bare minimum." For example, a Turing machine has very few possible instructions: it

---

[*] There is another way for the machine to execute high-level language programs without compiling: via an interpreter. But while a compiler translates code into standalone machine language code that is easily executable, a program written with an interpreter *always* requires the interpreter in order to run.

can read a symbol, write a symbol, halt the machine, and travel left or right. Yet that is sufficient to work through any mathematically computable problem.

Suppose we built a small computer with a handful of addresses for memory cells and a single accumulator. Like a Turing machine, we need to be able to stop the machine; let's call this a *halt* instruction. We also have to be able to read values from the memory cells; to do this, we will load them into our accumulator. Let's call this a *load* instruction. A Turing machine is able to not only read, but write symbols; we'll need something equivalent for our small computer. So, let's create a *store* instruction that takes the contents of the accumulator and copies those contents to some specified memory cell. All three of these instruction codes (also called opcodes), mind you, aren't stored in the computer as English words, but as binary numbers. Furthermore, to have some coding flexibility, we'll have (almost) every instruction code contain the address of the memory cell in RAM that it refers to. So, for instance, a load instruction won't just tell the computer it's time to load a value into the accumulator, it will also tell the computer *where in the memory that value can be found*, courtesy of that value's address. (The halt instruction, though, won't need an address.) That way, there doesn't need to be separate RAM for the instruction codes and the data, as in the old Harvard architecture model; both instruction codes and data can reside together in the same area of memory.

Right now, all our computer can do is load preset values from memory cells into its accumulator and copy them into other memory cells. But recall that the prime function of an accumulator isn't simply to be a holding pen for data; rather, it is to perform arithmetical operations on data, such as addition and subtraction (which, recall, can be performed by bit inverting the number to be subtracted, and then adding that number to the value present in the accumulator). With that in mind, let's create two more instruction codes: *add* and *subtract*, which will add or subtract a value stored in a memory cell to the contents of the accumulator. Now we have a computer that can work through simple arithmetic problems.[*]

There's another issue, however. Right now, our computer runs through the addresses sequentially, like a speeding train with broken brakes. An (unconditional) *jump* instruction gives us the flexibility to, quite literally, jump to a different address when the instruction's encountered by the computer, instantly dropping the speeding train at a different point on the tracks. A *call* instruction does the same as jump, but, before it jumps, it logs the previous address for future use; and a *return* instruction returns to that previous address.

---

[*] When working with numbers larger than a single byte, there are some additional instruction codes needed for carrying and borrowing, but we will ignore those complications here.

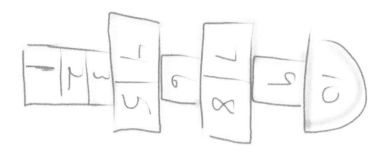

Even better than an unconditional jump, however, would be a *conditional jump* instruction, which is akin to playing a game of hopscotch: the jump would take effect only if a condition in the program is met. A conditional jump thus permits a common staple of computer programs, finite looping, first spelled out by Ada Lovelace in her *Notes*: "Both for brevity and for distinctness, a *recurring group* is called a *cycle*. A *cycle* of operations, then, must be understood to signify any *set of operations* which is repeated *more than once*. It is equally a *cycle*, whether it be repeated *twice* only, or an indefinite number of times...." Conditional jumps are the building blocks of the If... (also called If...Then) and For... (also called For...Next) statements of high-level languages like BASIC. Conditional jumps also open up the possibility of our simple computer multiplying or dividing numbers, all without the need for a *multiply* or *divide* opcode. We'll explore how later on.

There is one more thing we need to mention: the use of language. Instead of writing out in full the opcodes (operation codes) and their associated operands (arguments)— such as "store the value from the accumulator into address [xxx]"— it makes more sense to abbreviate using mnemonics. So "store the value from the accumulator into address [xx]" might be abbreviated this way:

> STO [xx],A   ; Store value into address

where the "STO" is the mnemonic for store, the [xx] represents the destination address, the "A" represents the accumulator, and the semicolon signifies the start of a comment (a for-humans-only English explanation ignored by the compiler).* In general, the format for mnemonics is as follows:

> *Mnemonic* [destination],[source]   ; Comment

But since, in this example, our simple computer is a single-accumulator-based machine, we can drop the "A" from the instruction because it's implied, resulting in the simpler instruction

---

* In most programs going forward, we won't require that comments begin with semicolons.

STO [xx]   ; Store value into address

But how can we input data into the computer (such as numbers to add), or receive output (such as the results of calculations)? Input and output can be controlled by the *in* and *out* instructions, respectively. The *in* instruction reads a value directly in to the accumulator, while the *out* instruction writes the contents of the accumulator to output.

There are several more commonly used mnemonics that need to be addressed here: PUSH and POP. These instructions are used with "stacks." Imagine making a house of cards: you carefully build up the house, from the bottom up, crossing your fingers that a small gust of wind doesn't flatten all of your hard work. Suppose you finish most of the structure, but then are dissatisfied with a portion of it: how can you make changes? Well, you can't pull cards from the bottom— that will cause the house to collapse. Instead, you'll have to carefully remove cards from the top.

This house of cards is a metaphor for a stack, a way of allocating memory (usually to capture and save the state of some program that has been interrupted but which will be returned to later). With a stack, we can only take off the object last put on, and vice versa; stacking is also commonly referred to as LIFO, or last-in-first-out. If we want to put an element on top of the stack (place a new card on top of the house of cards), we *push* it on; if we want to take off the topmost element of the stack (remove the card at the top of the house), we *pop* it off. When using your computer, you may have encountered the error "stack overflow"; essentially, this means that the stack has grown so large, it has used up its allocated memory.[*]

By the way: the aforementioned call instruction, which jumps to a new address but not before saving its prior address, pushes this address on the top of the stack. The return instruction pops this prior address back off.

paper computer design implications ♦♦♦♦♦♦♦♦♦♦♦♦♦♦♦♦♦♦♦♦♦♦♦♦♦♦

All three paper computers have most of the core set of opcodes described above: load, store, halt, add, subtract, in, out, unconditional jump (the CARDIAC technically uses the call/return opcodes instead of the unconditional jump, though we will nevertheless for convenience term it an unconditional jump), and conditional jump. Opcodes can always be abbreviated with mnemonics; from there, the differences between the paper computers abound. Also, stacks can be coded (and are implicitly a part of the CARDIAC's operation, courtesy of its unconditional jump), but push and pop are not explicitly part of the instruction set of any of the paper computers. And all three paper computers have one key restriction: the number of machine language statements in any program cannot exceed one hundred— since all three machines are limited to one hundred memory cells.

---

[*] A related concept is the "queue," which is FIFO: first-in-first-out.

As Charles Petzold playfully observes, writing in machine language is "like eating with a toothpick. The bites are so small and the process so laborious that dinner takes forever." But machine language programming allows us to be as miserly as possible about preserving a computer's most important finite resource: memory.

Assembly language is a step up from machine language. Rather than code zeros and ones, or even mnemonics that require you to explicitly set the memory locations for data via addresses in the instructions, we can use "labels," permitting the computer— automatically, behind the scenes— to set and manage the memory locations. For instance,

$$STO\ [xx]$$

might shortcut to something like this

$$STO\ [operand]$$

where [operand] is just that: a label pointing to the address of where you'll be storing the contents of the accumulator, or a reference to some other command.

Labels are used to jump to the start of "subroutines," or lines of code that are grouped together to (perhaps) be utilized more than once in a program. For example, the pseudo-assembly program

Pseudo-Assembly Program 1-1: Jump and Store

LABEL	MNEMONIC	OPERAND	COMMENTS
	JMP	start	Jump to label start.
	STO	A	Store value into variable a.
start	STO	B	Store value into variable b.
A	DAT		Define address for variable a.
B	DAT		Define address for variable b.
	HLT		Halt the machine.

will only store the value of variable b into an address, skipping over storing variable a— thanks to a jump opcode, which shuttles the program to the "start" label. The DAT mnemonic sets aside an address for the value stored in the variable. Effectively, DAT initializes a variable. We'll also be able to use DAT to assign the value of a variable to a constant.

You'll notice that assembly language allows us to work directly with variables, whereas machine language does not. Variables provide us with a convenient shorthand for storing data values: declare a variable by a name, assign a value to it, and that value can be retrieved later by simply calling the variable. The assembler, without prompting, chooses a memory location and, thus, an address in which to store the variable's contents.*

Writing code in assembly language is significantly easier than in machine language, but an assembler is required to translate the purely symbolic low-level assembly code (the "source-code file") into the ones and zeros of machine language (the "executable file")— which is all a computer can directly understand.

Nearly all high-level languages allow you to declare "arrays," which group together variables of the same type. Arrays organize a mass of data by indexing each element. For example, suppose you were a teacher with a class of thirty students, and you just scored a test taken by the entire class. It would be counterproductive to store each student's test score in a uniquely named integer variable, such as JakeScore, JenScore, BobScore, etc. Instead, an array called (let's say) TestScore could house all thirty scores by an index: so, TestScore[1] could contain the value of Jake's test score, TestScore[2] might contain Jen's test score, and so on. That way, we've compressed thirty separate variables into a single one. Furthermore, the array's indices could correspond in ascending order with the alphabetical order of the last names of the students; that way, the data set is organized in a systematic, easy-to-retrieve manner.

Some high-level languages, such as C,† also allow for "pointers." A pointer is a special kind of variable that holds the address or location of a variable besides itself. A pointer quite literally points to a location in memory of some other variable. When we obtain the data stored in a pointed-to address, that's called "dereferencing" a pointer, a form of "indirect access" to the value stored in the address (since we're not accessing the variable directly, only the contents contained in the variable's address courtesy of a pointer). Pointers can even point to other pointers, which is referred to as "double indirection." Pointers can also be utilized in the form of "linked lists" to help make arrays easier to manage, as well as with stacks in the form of "stack pointers," which reference the top of the stacks.

---

* We won't be venturing into variable types, such as numeric and string, in this guide. Assume henceforth that all data values are integers, sans the contents of the Instructo's storage locations (which, with its laissez-faire approach to memory, can contain any type of data without prior declaration).

† Created primarily by Dennis Ritchie at Bell Labs, C is only debatably a high-level language; others have perhaps more accurately referred to C as a "high-level assembler," because it offers large degree of low-level control (as does its object-oriented successor, C++, the "++" which, in C code, increments the "variable" C by one). Not only does C permit pointer declarations, it is also one of the few languages that has bit shifting operators built in: « for a left shift, and » for a right shift.

Finally, the manner in which programs are coded, regardless of language, varies from algorithm to algorithm and from programmer to programmer. Sometimes, programs employ "recursive" subroutines (or functions) which, rather dizzyingly, call themselves; recursive algorithms can be used to great effect when implementing iterative mathematical formulas, such as factorials,* and with techniques that search for data in arrays. In addition, a wide variety of sorting algorithms can be employed to manage and display data.

No matter how we approach coding, however, it's always best to keep in mind several of the dictums of Murphy's Law of Programming:

> Every nontrivial program contains bugs.
>
> *Corollary 1:* A program with more than 10 lines of code is nontrivial.
>
> *Corollary 2:* A program that's trivial contains no bugs.

Murphy's Law of Programming always is in force, especially as the length and complexity of your programs increase. Anyone who tells you otherwise is either lying or a fool.

paper computer design implications

After mastering paper computer machine language code, we will write paper computer assembly code; subroutines are a common stable of paper computer programming. Also, although paper computers don't allow for arrays or pointers directly, pointers can be supported indirectly through clever programming.

The next two sections will focus on programming the CARDIAC, LMC, and Instructo paper computers.

---

* E.g., the factorial of 6 is equal to $6 \times 5 \times 4 \times 3 \times 2 \times 1 = 720$. Factorials are usually employed in probability counting problems.

SECTION 2.

the cardiac and the little man computer

↳ Run CARDIAC/LMC Machine Language Programs by Using
The Pink Paper Computer (see Section 5)
Cardiac-Little Man Emulator (see Section 6)
Emulators Available Online
Original or Replica CARDIAC Hardware
Or Your Mind

↳ Run CARDIAC/LMC Assembly Language Programs by Using
Cardiac-Little Man Assembler (see Section 7)
Emulators Available Online
Or Your Mind

It is a high school math classroom in the late 1960s. The room is dry and cramped and stuffy and dark despite the propped-open screen-less windows and flickering fluorescents. Visible chalk particles float in the air as a middle-aged, balding, bespectacled, frumpy, slightly hunched caricature of a man bedecked with a mustache and wearing a starched white half-tucked-in shirt with a pocket-protector stands facing an unwashed chalkboard with unclapped erasers resting nearby on a chalk tray. The man reaches for a stick of white chalk, leans over, presses the stick firmly to the board, the pressure causing the tips of his fingers to slightly swell and turn pink— and then stops.

"Instead of this usual stuff, how would you like to learn about computers today?"

A hush falls over the classroom. Students who were chewing gum, shaking to the beat of a song playing only in their minds while fiddling with their unkempt long hair, glance upward toward the teacher. Other students stop, however momentarily, their low-volume steady-state whispering behind the teacher's back.

"I know we don't have a computer in here, or in the school," he says. The air quickly leaves the room as audible guffaws in sotto voce fill the space instead.

"Shush. But earlier today, a package arrived from the phone company." He holds up a large cream-colored box with turquoise lettering on it; on the side, students in the front row can just barely make out the word "cardiac" splayed near the top of the box in a kind of pointillistic spread, with a big Bell Labs logo affixed to the bottom.

The teacher reaches into the box and pulls out a handful of glossy turquoise-and-cream die-cut cardboard packages, each of which unfolds to reveal little light-blue pamphlets— CARDIAC manuals— and thick, card stock sheets with symbols printed on them, some assembly required.

"I want you guys to get in groups of two or three. If you want, just push your desks next to each other. I'll come around and hand one of these CARDIAC computer kits to each group to assemble."

No one moves.

"Come on, guys, let's go. No one's working alone on this."

The squeaks of metal desk legs on linoleum-tile sound like a concert of reluctance, as the students— some barely lifting their behinds off their seats— drag their desks slowly around the room.

The teacher, starting intently with eyebrows furrowed, walks the perimeter, places a kit down next to each cluster of students, and glides to the front of the room.

"Okay. Good. Let's get started. Open the fold and look at the page with the heading, 'CARDIAC ASSEMBLY INSTRUCTIONS.' It has the diagrams printed on it. Got it?[*]

"Okay. Good. Looking at these instructions, there's a bit to do. Check to make sure you have the five 'bugs'— they are the cutouts of ladybugs and are going to point to which 'address' we're working on."

A thin doe-eyed girl with shoulder-length golden braided locks and bangs raises her hand.

"Yes Patti?"

"What's the point of all this? I mean, why are we doing this?" A wave of snickering ripples through the room.

"This CARDIAC," the teacher says, "this kit that we received from the phone company, is going to help us learn about computers. It is not a computer."

"It's obviously not a computer," Patti interjects, dripping sarcasm.

"Right, it's not."

---

[*] http://www.kylem.net/hardware/cardiac/CARDIAC_scans.pdf

"So what the heck is it?"

"Language, please. The CARDIAC stands for a 'computer illustrative aid to computation.' We're going to be doing the calculating part, and using the CARDIAC to 'process' a program like a real computer would process it."

Visibly frustrated, Patti sighs loudly, blowing the bangs from her eyes. "Its grooves don't look groovy, but let's do it."

"Okay. Good. You can try putting the ladybug into one of the holes on the memory board, but you'll have to punch the holes out first. Now, put the bugs aside, since we'll only need one later on, and punch out all of the little holes in the memory cells. You can clean them up later, just do it now, quickly.

"Then, fold the CARDIAC along the 'score marks'— they have the dashed lines; do you see them?— as it says to do in instruction three, and carefully fit the four function slides through the pre-cut slots." The sound of folding and popping and crumpling fills the air.

"What's an OP Code?" asks Megan, one of the few earnest students.

"It stands for 'operation code,' and I promise we'll get to that later. Right now, I want you guys to finish up the assembly by making sure the four slides can actually slide up and down in the cardboard. Got it? Okay. Good. And now— oh, wait, you're all going to need some glue to seal the bottom. Give me a second here, let me hand that out to you."

The teacher reaches underneath his desk and pulls out five containers of school glue, each adorned with bright orange caps and white labels that scream "WASHES OUT."

"Add just a little glue to the bottom of the CARDIAC to seal it, but don't get any on the slides or slots or the die-cut holes or anything! Got it? We are almost ready to program the thing! I've already assembled the Vu-Graph CARDIAC, which will help me demonstrate to you how to run program. The left panel of your CARDIAC houses its CPU, its central processing unit where it 'thinks,' while the right panel— with the holes— functions as its memory. But first, let me fetch the projector because I want to show you a film that came with the CARDIAC, called 'The Thinking Machines.'"

tales of cardiac nostalgia

This scene, circa 1969,[*] could have just as easily been set in the middle 1970s or even the early 1980s, where a CARDIAC demo might set the stage for a

---

[*] Although I of course took some artistic license, the preceding scene is a mélange of the

class's first visit to a newly stocked TRS-80 or Apple II computer lab, akin to a driver's education class preparing a group of gangly greenhorns for when the rubber meets the road.

There are number of online forums, including an active Google+ Community replete with CARDIAC videos and emulators,[*] with postings from the (mostly) middle-aged describing their recollections of the CARDIAC— from both their high school and middle school years— and of people a generation younger recounting tales their parents told them of first encounters with a computer constructed of cardboard. Inevitably, talk of the CARDIAC triggers a flood of nostalgia not just about the "pulp device," but also of programming early punched-card machines in FORTRAN, or of using time-sharing terminals in BASIC or COBOL. Thinking about those first CARDIAC encounters grants folks permission to be young again— and instantaneously sharing their computing experiences (and obsessions) with the many like-minded became feasible beginning in the mid-nineties, courtesy of the medium of the internet.[†]

"I remember a sixth grade assembly being introduced to these [CARDIACs]," one post read. "It was my first lesson in computers. Not everybody got it but my friend and I looked at each other and smiled— we got it right away."

"This reminds me of when we used an IBM keypunch machine to run FORTRAN programs. Also, we used the CARDIAC to write games, like Nim," another post read.

"We used these things back in 1980! It didn't make sense to some of my friends when I ran the slides up and down and ran a program for them. So then I got new friends," read a third post.

Other people immediately connected the CARDIAC with the LMC: "It really reminds me of the Little Man." Still others satirized about what would happen if we were ruled by "5,000-instruction set cardboard-computer robots" someday in the future. "Will they be our benevolent overlords? I hope so."

An early bulletin board thread from 1994, launched with a freewheeling discussion of the CARDIAC— with the then-23-year-old writer working hard to remember the details of the paper computer without the benefit of the infinite resources of an internet search we have today— notes that "I have a vague feeling this teaching aid was put out by somebody respectable, like Bell Labs. Anyway, my fifth/sixth grade teacher used this to prepare us for the future...." Quickly, the thread jumps to related computer engineering topics, like logic gates, as well as descriptions of other contemporaneous computer

---

many anecdotes I've read and heard about people's first introduction to the CARDIAC.

[*] https://plus.google.com/communities/105401067945055411018

[†] Which also helped to balkanize thought and experience as well.

kits, such as the Digi-Comp I (a mechanical digital computer sold in the sixties), the Geniac, and the Simon.*

the cardiac manual

The CARDIAC manual is a slim, 53-page, Preface-plus-sixteen-section, prodigiously illustrated† booklet with a light-blue cover that was slipped inside each CARDIAC device. Just as much as the CARDIAC itself, the manual is sui generis: the half-century-old booklet, which went through numerous printings, with its simple examples, cogent writing, and bold diagrams, at times reads like an urtext of the later *Dummies* series of computer books, among other introductory texts. But it's far from perfect and is most certainly a product of its time. What follows is a précis of the manual.

The Preface is perhaps its weakest part. With a cursory Bell Labs-centric discussion of the history and utility of computers, much is left out or glossed over; why should we care that Bell Labs' technical personnel devotes ever-increasing time to programming computers? And, as mentioned earlier, who is the signatory: G.I.R.?

The short first section makes clear that the CARDIAC is not a computer, but merely an aid to computation designed to help illustrate the functionality of computers for these "fast-moving times." It is, essentially, a simulacrum, a tangible metaphor for a computer.

The second section lays out von Neumann architecture. Instead of the control unit that directs the ALU, the authors present the equivalent "program unit" that directs the computer's processing of inputs, outputs, calls to memory, and accumulator operation. We're not quite ready yet to use the CARDIAC; instead, how this layout for a "fairly simple computer," abbreviated SIMCO, works is the focus. Specifically, how SIMCO could process a generalized algorithm for finding the sum of two three-digit numbers is detailed.

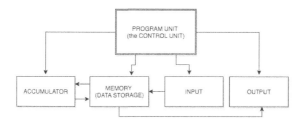

Fig. No. 6. Layout of SIMCO.

---

* http://www.dorje.com/netstuff/folklore/toy.computers

† By an A. Barthelson. Further information about the artist or artwork is unavailable.

The third section describes how computers use pulses to communicate with each other in binary form, as well as detailing the input and output devices available at the time, such as punched cards and printers.

The fourth section explains the function of action diagrams, which illustrate the data flow of a program, as well as flowcharts, which describe a program's step-by-step procedures. One of the example flowcharts presented, for repairing a flat tire, is flat-out sexist. The first question, wrapped inside a diamond, reads as follows: "Are you a girl?" If you answer yes, then you're taken to a rectangle with the instructions, "Look forlorn and helpless until some unsuspecting Galahad stops and fixes flat"; if you answer no, then you are instructed to hide if a girl is with you so that she can "look forlorn and helpless," etc. Quite demeaning suggestions, and ones that would land at least one mathematics teacher in a bit of public trouble.

On May 19, 1975, the *Sydney Morning Herald*— the same newspaper that would review the Instructo five years later— featured an article on classroom computers by the male head of the math department at Marrickville Girls' High School just outside of Sydney, Australia. In the article, the flat-tire flowchart is lifted wholesale from the CARDIAC manual and presented both humorously and innocuously.

But not everyone took kindly to the piece. A woman named Anne Dupree quickly shot off an angry Letter to the Editor, which was published on May 21. "SIR, The article on classroom computers... was interesting indeed, as an example of male chauvinist piggery," she begins. After methodically describing the sexist wording, she rightly concludes, "Patronising humour of this kind does nothing to educate girls."[*] It was such moment of pride for Dupree— a multi-award-winning lifelong journalist who had once years before worked for the *Herald*— that the *Herald* letter is described in the second paragraph of her obituary.[†]

Dupree's point is well taken: Why discourage girls from programming computers? Later, in the sixth section of the CARDIAC manual, there's another chauvinistic swipe, when the interpretations for the numbers 38-24-36 run the gamut from football signals to "the dimensions of a curvaceous actress" (who is drawn on the page, well, curvaceously).[‡] Dismissing such sexist implications as mere poppycock does a disservice to female students who may

---

[*] https://news.google.com/newspapers?nid=1301&dat=19750521&id=SfpjAAAAIBAJ&sjid=YeYDAAAAIBAJ&pg=5231,6790145&hl=en

[†] http://www.smh.com.au/comment/obituaries/war-bride-world-traveller-and-awarded-journalist-20090405-9t6u.html

[‡] Recall a similar scene in the film "The Thinking ??? Machines," with a female of the exact same dimensions. Note that the only individual who is credited contributing to both the film and the manual (he is listed as a writer on each) is Saul Fingerman.

have had an interest in computing, only to be subtly turned away by the language of the manual and institutional chauvinism it enabled male classroom teachers to get away with. Puzzling, considering the stories of Ada Lovelace, Charles Babbage's collaborator, and Grace Hopper, who, by the 1960s, was already recognized as an intellectual pioneer by helping to develop COBOL and creating the first compiler, which translates statements written in algorithmic programming languages into the machine language of computers. Lovelace and Hopper could have, and should have, been held up as exemplars of women who coded, let alone the many women who were coding computers from ENIAC onward.

Continuing to run through the manual, the fifth section introduces the operation codes* of the SIMCO— which, as we'll see in subsequent pages, are identical to those of the CARDIAC— as well as explaining the layout of the one hundred memory addresses. And the basic idea of von Neumann architecture, that program instructions and data are treated the same way, is addressed in a short subsection entitled "Instruction Words and Data Words Are Look-Alikes": "This not only makes for greater economy," the manual emphasizes, "but also means that, as a computer is proceeding through a problem, *it can process its own* instructions." Thus, SIMCO (as well as CARDIAC) is a stored-program computer with von Neumann architecture.

The layout for the stored-program computer is detailed in the sixth section. It is here that we see the recognizable elements of a von Neumann machine: besides simply input, output, control unit, and accumulator, there is an instruction register— with an opcode and operand— sitting at the ready, as well as a program counter and cells stacked in memory. It is also here that the dimensions of a "curvaceous actress" are utilized as an example of how SIMCO interprets the operand as either an instruction word or a data word: it's all about context. The control unit keeps everything humming along smoothly, incrementing or decrementing the program counter as needed as well as fetching instruction words from memory cells to place into the instruction register.

It is only when we arrive at the seventh section of the manual that the CARDIAC is explicitly introduced— by building a short bridge between the theoretical SIMCO construct and the literal cardboard computer kit (hopefully) sitting all assembled right in front of you. Inputs and outputs are handled

---

* Although the CARDIAC manual refers to these as "OP Codes," the language has evolved over the past half-century; we will continue to use the modern, concatenated word "opcode" as the correct term.

on strips of paper, which slide in and out of the device; memory consists of one hundred three-digit decimal words, written onto and erased from memory cells via pencil; steps of arithmetic are completed in the accumulator, replete with a slide expressing the sign (positive or negative), and restricted to four digits should there be overflow past 999; a pointer, with a ladybug head, functions as the program counter, fitting into any of the circular slots punched out of each of the one hundred memory cells; and an instruction register relays the current opcode and operand. SIMCO's control unit becomes CARDIAC's control unit, of which you are the star: since you'll need to write, do arithmetic, push and pull slides, and move bugs.

The remainder of the manual— sections eight to sixteen— focus on software (writing code) rather than hardware (the physical device). From the power and pitfalls of loops in the ninth and tenth sections, to examples of multiplication in the eleventh, to shifting (decimal) digits in the twelfth, to subroutines in the fourteenth, to the development of a game of Nim in the fifteenth,[*] there is much detail to parse through. We will create our own sample programs illustrating these concepts shortly.

Before we leave the manual, however, there are two more sections that should be mentioned: the thirteenth and sixteenth. The former section discusses loading programs by bootstrapping, instead of by simply penciling the programs right into arbitrary memory cells, while the latter section elucidates the differences between assemblers and compilers, presenting an example CARDIAC assembly language program. We will explore the concepts in these two sections in great detail later.

The manual is cogent, clear, and complete, and holds up (mostly) well fifty years after its publication, largely because computer architecture, at its most basic, hasn't changed— and the CARDIAC purports only to model the basics of computers, after all. Besides the occasional sexist and otherwise antiquated language, the most trenchant criticism that can be levied at the manual is the one of "radical novelty," or a development that brings a change so great it results in a "sharp discontinuity" from previously lived experience and common sense that the computer scientist E. W. Dijkstra famously spoke about in his "On the Cruelty of Really Teaching Computing Science" (1988).[†] Computers and computing, he argued, represent a radical novelty (or, more precisely, two such novelties).

---

[*] Nim, a two-player strategy game in which each player must remove at least object per turn until none remain, was a popular program loaded into the primitive toy computers of the time. For example, Edmund Berkeley's Berkeley Enterprises manufactured the Nim Machine, where players tested their mettle against an electronic opponent. Even earlier, there was the Nimrod, an unfortunately named digital computer expressly designed to play Nim, featured at the 1951 Festival of Britain.

[†] https://www.cs.utexas.edu/users/EWD/transcriptions/EWD10xx/EWD1036.html

Coping with radical novelty requires an orthogonal method. One must consider one's own past, the experiences collected, and the habits formed in it as an unfortunate accident of history, and one has to approach the radical novelty with a blank mind, consciously refusing to try to link it with what is already familiar, because the familiar is hopelessly inadequate. One has, with initially a kind of split personality, to come to grips with a radical novelty as a dissociated topic in its own right. Coming to grips with a radical novelty amounts to creating and learning a new foreign language that cannot be translated into one's mother tongue.

By making use of "metaphors and analogies we try to link the new to the old, the novel to the familiar," but this won't do, since the radical novelty represents not a difference in degree but a difference in kind.

Yet populated throughout the CARDIAC manual we find appeals to the old and familiar to help explain the computer: arithmetic problems solved on a chalkboard, descriptions of the "alphabet" of machine communication, the repairing of flat tires, telephone operators setting up calls, football plays, dimensions of "curvaceous" actresses, a magic "nines" number trick, and even a program counter's physical manifestation: a picture of a ladybug on cardboard. Worse still might be the LMC: a homunculus bouncing around inside a computer anthropomorphizing its bits and bytes. If a human really did live inside your computer, now *that* would be radical novelty!

Truth be told, this guide is also guilty of using old and familiar metaphors to introduce and explain concepts, as are whole hosts of books and resources on computers and computer programming. Nonetheless, we press on.

a quick word on the vu-graph

The for-teachers-only Vu-Graph CARDIAC, packaged in a large envelope reading "TEACHER'S GUIDE," is, in effect, a transparent CARDIAC (a clear, thick transparency, with black printed text and diagrams), complete with input and output slides and all the other accouterments that go along with the cardboard computer— with several notable exceptions. First, there are three columns of memory cells missing, leaving only fifty memory cells. Second, two opcodes— 3 (TAC), the conditional jump, and 4 (SFT), the digits' left/right shift— are "disabled." As the assembly instructions sheet, under the "TEACHER: PLEASE NOTE" heading, states, "[T]hese adjustments limit your use of the Vu-Graph in only a few of the more advanced programs set forth in the CARDIAC student manual. In most cases where Op Codes 3 and 4 are to be used, or the suggested memory column is missing, you can adjust by taking the direction from the students and by re-marking memory addresses on the Vu-Graph."

the central conceit of the little man computer

In the LMC, a little man in a cloistered mailroom performs all of the computing tasks, which turn out to be quite a bit of physical labor; this labor is classified as either the *fetch* portion of a cycle, or the *execute* portion. He "fetches" by reading the program counter and running to the address listed on the program counter. He then "executes" by thinking about what, precisely, the instructions at that particular address are telling him to do, and then actually doing them.

The little man is kept busy in the "execute" cycle, gathering data from the inbox (the area housing input), manipulating values in the calculator (i.e., the accumulator) to perform some sort of arithmetical calculation, looking up values in mailboxes (i.e., memory cells), or printing data to the outbox (the area receiving output). The little man receives inputs on slips of paper he retrieves from the inbox; outputs are printed on slips of paper he slides into the outbox. Once the execute cycle is complete, the little man increments the program counter and restarts the cycle anew.

The little man is a native decimal speaker; he can't comprehend binary. When he sees a three-digit value stored in a mailbox, he usually interprets the first digit as a particular instruction to perform, and the latter two digits as reference to an associated mailbox.[*] There are one hundred mailboxes, each with the capacity to hold a three-digit number ranging from 000 to 999. As the article "Little Man Computer— An Introduction to Computer Architecture and Operation"[†] makes clear, "Keep in mind that, whilst the LMC works with decimal numbers, real computers use binary and the width of busses is measured in bits. Our choice of bus widths means that we cannot have more than 10 different operations," since the width of the address bus is two (two digits required for each address), leaving one digit for instructions.

---

[*] There are several exceptions to this format, as we will see shortly when examining instruction sets.

[†] https://ecs.victoria.ac.nz/foswiki/pub/Courses/ENGR101_2016T1/LectureSchedule/LMC_descrip.pdf

And, unlike the CARDIAC, all memory in the LMC is volatile; the contents of any mailbox can not only be read, but also written onto.

our approach to coding

In each subsection henceforth, as we explore the nuts and bolts of programming, we will toggle between CARDIAC and LMC syntax. Both paper computers are sufficiently similar that we can present them side by side, comparing and contrasting their educative approaches along the way.[*]

the opcodes

The instruction sets for the CARDIAC and LMC overlap, but there are differences. Here is the instruction set for the CARDIAC:

CARDIAC Instruction Set

OPCODE	MNEMONIC	DESCRIPTION
0	INP	Reads a single input into a designated memory cell.
1	CLA	Clears the accumulator of data; inserts value present in a designated memory cell into the accumulator.
2	ADD	Adds the value present in a designed memory cell into the value already found in the accumulator.
3	TAC	Tests the accumulator's value; if the value is positive or zero, then increment the program counter; if the value is negative, then jump to a designed memory cell.
4	SFT	Shifts accumulator value x places left (x - first digit of operand) and y places right (y - second digit of operand), replacing shifted digits with zeros.
5	OUT	Outputs the contents of a designated memory cell.
6	STO	Stores the accumulator's value into a designated memory cell.
7	SUB	Subtracts the value present in a designed memory cell from the value already located in the accumulator.
8	JMP	Stores the "current" memory cell's address + 1 into memory cell 99 (so the memory location can later be re-referenced), then performs an unconditional jump to the instruction in the designated memory cell.
9	HRS	Halts the machine, and reset the program counter to zero.

---

[*] You might be wondering why we've ignored the Little Man Computer manual. It's because it was only much later, in 1979, that Madnick drafted anything that might even be loosely deemed an LMC manual, called "Understanding the Computer (Little Man Computer)"; written as a working paper, it turned into an unpublished manuscript that was translated, years later, into French and Spanish. It's the closest Madnick, who is an incredibly prolific writer, ever got to publishing a primer on the LMC. (The sheer volume of his academic publications is astounding; beyond that, he even penned three *unpublished* textbooks. See his curriculum vitae at http://web.mit.edu/smadnick/www/Resume/Publications.htm)

The CARDIAC's JMP mnemonic is a bit tricky; essentially, it is a call/return instruction. For example, if the jump instruction is called in memory cell 60, then the following is stored into memory cell 99: 861 (an unconditional jump to memory cell 61). This "save state," which stores a return address, is useful when accessing subroutines. Each time the CARDIAC starts up, memory cell 99 by default contains the contents 800 (or 8--, which is, in effect, the same thing)— an unconditional jump to memory cell 00, the contents of which are unchanged by the jump (since the "current" memory cell's address + 1 = 00).

The LMC has no commensurate "save state" instruction. Here is the LMC's instruction set:

Little Man Computer Instruction Set

OPCODE/ NUMERIC CODE	MNEMONIC	DESCRIPTION
1	ADD	Adds the contents of a designated mailbox to the value present in the calculator (accumulator).
2	SUB	Subtracts the contents of a designated mailbox from the value present in the calculator (accumulator).
3	STA	Stores the contents of the calculator (accumulator) into a designated mailbox.
5	LDA	Clears the accumulator of data; inserts value present in a designated mailbox into the calculator (accumulator).
6	BRA	Branch unconditionally to the instruction in a designated mailbox.
7	BRZ	Tests calculator (accumulator) contents: if the contents equal zero, then jump to a designated mailbox; if they equal anything else, then increment the program counter.
8	BRP	Tests calculator (accumulator) contents: if the contents equal zero or are positive, then jump to a designated mailbox; if they equal anything else, then increment the program counter.
901	INP	Reads a single input into the calculator (accumulator).
902	OUT	Reads the calculator's (accumulator's) value and copies it to the outbox.
000	HLT/COB	Halt the machine (also called a "coffee break").

There are a number of subtle differences in the two instruction sets that are important to note right away. First, the LMC has no "bit" shift mnemonic (it can, however, be programmed); but the LMC does have an additional conditional jump option not present in the CARDIAC (BRZ). Also, the LMC's input and output instructions have both an opcode and operand already built in: 901 and 902, respectively. Here's why: unlike with the CARDIAC, where the input can be read *directly* to a specific memory cell or the output read *directly* from a specific memory cell, LMC inputs and outputs are mediated by the accumulator (also called the calculator). The path of data traveling from outside the machine, to inside into the accumulator or memory, and then back out again, is where the majority of differences between programming the CARDIAC and LMC lie.

At first you might think the CARDIAC's and LMC's "processor architectures" belong into the class called a Minimal Instruction Set Computer (MISC). Some early digital computers were Minimal Instruction Set Computers, mostly because they had very few opcodes. Most MISCs are stack-based machines, however, and neither the CARDIAC nor the LMC has the opcodes necessary to directly manipulate stacks. Perhaps the CARDIAC and LMC architecture can be more accurately termed Reduced Instruction Set Computing (RISC).

building a calculator program

Pascal and Leibniz and Babbage, among many others, wanted to build efficient automated calculating machines. Let's begin our programming exercises by, from the ground up, coding a functional automated calculating machine— in essence, a calculator— for both the CARDIAC and the LMC.

A quick reminder, before you get too excited: just because we write a calculator program doesn't mean the CARDIAC or LMC has *become* a calculator— since, after all, we always need to complete all of a paper computer program's intermediate calculations by hand. And a quick note: recall that, when speaking of the LMC, the accumulator is usually referred to as a "calculator"; please use context clues to not to confuse the *calculator* program we're about to build with the frequently mentioned *calculator* (accumulator) of the LMC.

Both the CARDIAC and the LMC have ADD mnemonics; in addition, the CARDIAC has the CLA mnemonic, which we'll also need to make use of. Although it might be tempting to begin our program at the first available memory cell (01 for the CARDIAC, because memory cell 00 has 001 hardwired into it; and 00 for the LMC), we want to leave some space in memory in case we need to insert code later (the CARDIAC manual calls this providing "elbow room"). Below is our first attempt at writing a super-simple calculator program (for the CARDIAC and LMC), which can complete only one arithmetic operation: take two numbers from input, add them, and then output the result.

CARDIAC Program No. 2-1: Add Two Numbers

ADDRESS	CONTENTS	MNEMONIC	OPCODE	COMMENTS
17	090	INP	90	Read the input card into cell 90.
18	091	INP	91	Read the input card into cell 91.
19	190	CLA	90	Clear the accumulator, setting its contents to be the data in cell 90.
20	291	ADD	91	Add contents of cell 91 to the accumulator.
21	692	STO	92	Store the contents of the accumulator into cell 92.
22	592	OUT	92	Output the contents of cell 92.
23	900	HRS	00	Halt and reset.

Before running this program, you'll need to do two things: (1) Place the two numbers you wish to add on the input card, and (2) set the program counter (the "bug") to the memory cell designated by the first address of code shown, namely, cell 17. Henceforth, we won't explicitly state the initial value for the program counter— unless otherwise directed, set the program counter to the first memory cell address of the program.

Run the program. If you wrote 10 and 20 on your input card, for example, the output card would show 30. Note that because the accumulator only has space for a four-digit number— it is shown in a rectangular four-square arrangement on the face of the CARDIAC, one square per digit— any sum exceeding 999 is classified as "overflow."

Let's now rewrite the algorithm for the LMC.

LMC Program No. 2-1: Add Two Numbers

MAILBOX	CONTENTS	MNEMONIC	OPERAND	COMMENTS
17	901	INP		Reads the first number from the inbox into the calculator.
18	390	STO	90	Stores the value in the calculator into mailbox 90.
19	901	INP		Reads the second number from the inbox into the calculator.
20	190	ADD	90	Adds the value in mailbox 90 to the value showing in the calculator (i.e., the second input).
21	902	OUT		Copies the value contained in the calculator to the outbox.
22	000	COB		Halt the machine.

The LMC program required one fewer line of code than its CARDIAC counterpart because the accumulator (or calculator) acts as a kind of "buffer," an input-to-memory-cell intermediary.

Coding the complementary programs for subtraction would read identical to those listed above, save for one difference: memory cell 20 would contain the instruction for subtraction, not addition. Alternatively, you could keep the same ADD mnemonics but enter in negative numbers instead of positive ones wherever you wanted to subtract. For now, though, we'll put aside subtraction and focus on how to make our calculator program add more than just two numbers at a time.

To add three numbers with the CARDIAC, we might write a program that looks like this:

CARDIAC Program No. 2-2: Add Three Numbers

ADDRESS	CONTENTS	MNEMONIC	OPERAND	COMMENTS
17	090	INP	90	Read the input card into cell 90.
18	091	INP	91	Read the input card into cell 91.
19	092	INP	92	Read the input card into cell 92.
20	190	CLA	90	Clear the accumulator, setting its contents to be the data in cell 90.
21	291	ADD	91	Add the contents of cell 91 to the accumulator.
22	292	ADD	92	Add the contents of cell 92 to the accumulator.
23	693	STO	93	Store the contents of the accumulator into cell 93.
24	593	OUT	93	Output the contents of cell 93.
25	900	HRS	00	Halt and reset.

And the equivalent LMC program might look this way:

LMC Program No. 2-2: Add Three Numbers

MAILBOX	CONTENTS	MNEMONIC	OPERAND	COMMENTS
17	901	INP		Reads the first number from the inbox into the calculator.
18	390	STO	90	Stores the value in the calculator into mailbox 90.
19	901	INP		Reads the second number from the inbox into the calculator.
20	391	STO	91	Stores the value in the calculator into mailbox 91.
21	901	INP		Reads the third number from the inbox into the calculator.
22	190	ADD	90	Adds the value in mailbox 90 to the value showing in the calculator.
23	191	ADD	91	Adds the value in mailbox 91 to the value showing in the calculator.
24	902	OUT		Copies the value contained in the calculator to the outbox.
25	000	COB		Halts the machine.

Although adding three numbers this way works fine, the problem for both versions is memory used: if you want to add four numbers, or five, or six, your program will grow longer and longer. Add enough numbers, and you'll run out of memory.

We'll need to use a loop to gather the input, no matter how many numbers. But if we use an *unconditional* loop, we'll quickly run into trouble— as can be seen by the following two programs (one for the CARDIAC, the other for the LMC):

CARDIAC Program No. 2-3: Add Numbers (Forever)

ADDRESS	CONTENTS	MNEMONIC	OPERAND	COMMENTS
17	091	INP	91	Read the first number into cell 91.
18	191	CLA	91	The contents of the accumulator are cleared, and then replaced with the contents of cell 91.
19	091	INP	91	Read the next number into cell 91.
20	291	ADD	91	Adds the value present in cell 91 to the contents present in the accumulator.
21	691	STO	91	Stores the contents of the accumulator back into cell 91.
22	591	OUT	91	Outputs the contents of cell 91.
23	819	JMP	19	Unconditional jump to cell 19.

Although this program will add together any number of numbers you place into the input while using only as much memory as the code for adding two numbers, the program will not stop— it's set to loop *forever*, repeatedly adding zeros to the sum when the input card is completely read (see the flowchart below). And that just will not do.

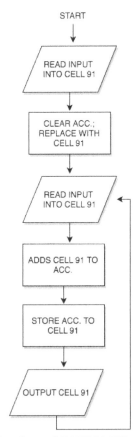

Fig. No. 7. Flowchart of CARDIAC Program No. 2-3.

Below is the equivalent LMC looping-forever program, which uses the same amount of memory as the CARDIAC version:

LMC Program No. 2-3: Add Numbers (Forever)

MAILBOX	CONTENTS	MNEMONIC	OPERAND	COMMENTS
17	901	INP		Reads the first number to be added into the calculator.
18	391	STO	91	Stores this first number into mailbox 91.
19	901	INP		Reads the next number to be added into the calculator.
20	191	ADD	91	Adds the value present in mailbox 91 with the number already in the calculator.
21	391	STO	91	Stores the contents of the calculator back into mailbox 91.
22	902	OUT		Copies the value contained in the calculator to the outbox.
23	619	BRA	19	Unconditional jump to mailbox 19.

Instead, let's write a CARDIAC program that, at least theoretically, adds together N numbers, where N is the absolute value of the first number the user writes on the input card (e.g., to add five numbers, the user first inputs -5), which is followed by the list of numbers he or she wishes to add. We're going to need to use a conditional jump (TAC), not the unconditional jump (JMP), otherwise the program will run forever. As you're examining the code below, notice that memory cell 00— which has the data 01 "hardwired" in— is leveraged as a loop counter, incrementing the value in memory cell 90 by 1 unit with each run of the loop. The loop ends only when the value of memory cell 90 is zero (or greater than zero), thanks to the TAC instruction.

CARDIAC Program No. 2-4: Add N Numbers

ADDRESS	CONTENTS	MNEMONIC	OPERAND	COMMENTS
17	090	INP	90	Read the first input into cell 90. In this program, this first input will keep track the number of numbers to be added (N).
18	091	INP	91	The second input is the first number of the list of numbers to be added; this first number is stored in cell 91.
19	191	CLA	91	The contents of the accumulator are cleared, and replaced with the contents of cell 91.
20	092	INP	92	The next number on the input card is stored into cell 92.
21	292	ADD	92	Adds the value present in cell 92 to the contents of the accumulator.
22	691	STO	91	Stores the contents of the accumulator back into cell 91.

MAILBOX	CONTENTS	MNEMONIC	OPERAND	
23	190	CLA	90	The contents of the accumulator are cleared, and then replaced with the contents of cell 90 (which keeps track of the number of iterations of the loop remaining).
24	200	ADD	00	The value in the accumulator is increased by one (which is the value in cell 00).
25	690	STO	90	The accumulator's contents are stored back into cell 90.
26	319	TAC	19	Tests the contents of the accumulator: if positive or zero, then simply increment the program counter; if negative, then jump to cell 19.
27	591	OUT	91	Output the contents of cell 91.
28	900	HRS	00	Halt and reset.

So, for example, if your inputs are -4, 10, 20, 30, and 40, then the program— after cycling through the loop enough times to add all the numbers (except the first one, -4) together— should output 100.[*] And, again, subtraction of numbers can be implemented here by either (carefully) retrofitting the code with the SUB mnemonic or simply inputting the numbers as negative that you wish to subtract.[†]

The LMC version of the program looks like this:

LMC Program No. 2-4: Add *N* Numbers

MAILBOX	CONTENTS	MNEMONIC	OPERAND	COMMENTS
17	901	INP		Reads a single value into the calculator. In this case, this value is the number of numbers to be added (*N*).
18	390	STO	90	Stores the contents of the calculator to mailbox 90.
19	901	INP		Reads the first number to be added into the calculator.
20	391	STO	91	Stores this first number into mailbox 91.
21	901	INP		Reads the next number to be added into the calculator.

---

[*] To be more precise, the inputs -3, 10, 20, 30, and 40 will also result in a correct output of 100— and an extra, unnecessary loop through the code (the program adds 0 to the running total in its final pass through); eliminating this extra iteration would result in the user instructions becoming even more convoluted: "The first number you write on the input card must be the negative whole number equaling one fewer than the number of items to add...."

[†] Purists might be given pause by the negative input in this program, but the CARDIAC manual never explicitly forbids it (some incarnations of the LMC do, however). In addition, note that a number of programs going forward employ negative words as data inside memory cells; as long as those memory cells aren't read *as instructions* but are only considered as data by either the CARDIAC or the LMC, the computer won't mind. But if data words are read as instructions, "the computer will go slightly insane," according to the CARDIAC manual.

22	392	STO	92	Stores this number into mailbox 92.
23	591	LDA	91	Clears the calculator; inserts the value present in mailbox 91 into the calculator.
24	192	ADD	92	Adds the value present in mailbox 92 to the number already in the calculator.
25	391	STO	91	Stores the contents of the calculator back into mailbox 91.
26	590	LDA	90	Clears the calculator; inserts value present in mailbox 90 into the calculator.
27	233	SUB	33	The value in the calculator is decreased by one (which is the data present in mailbox 33).
28	390	STO	90	Stores the contents of the calculator back into mailbox 90.
29	821	BRP	21	Tests the calculator's contents: if they equal zero or are positive, then jump to mailbox 21; if negative, then increment the program counter.
30	591	LDA	91	Clears the calculator; inserts the value present in mailbox 91 into the calculator.
31	902	OUT		Copies the value contained in the calculator to the outbox.
32	000	COB		Halt the machine.
33	001	DAT		Data, used to decrement counter.

Compared to the CARDIAC version of the program, there were a number of alternations necessary for the LMC beyond just taking into account the path of data inputs into the calculator (accumulator). First, instead of entering a negative number for the initial value of the counter, you'll need to enter a positive one. So, for instance, if you wanted to add the same four numbers— 10, 20, 30, and 40— you'd input the following: 4, 10, 20, 30, 40.[*] In addition, since the LMC doesn't have the data word 001 "hardwired" anywhere, as the CARDIAC does in memory cell 000, the LMC program inserts the data 001 into an arbitrary mailbox: mailbox 33 (notice the assembly language mnemonic DAT; in reality, it's not necessary here with the machine code, functioning merely as a placeholder); this data is used to decrement the loop counter.

But having to specify, in advance, the number of numbers to sum from the input card is suboptimal. It would be much more convenient if the program itself could determine when the entire set of inputs has been fed into the machine, and then print the output sum and stop the program. We'll need a terminating character— something that the computer spots which tells it to stop summing numbers. For the CARDIAC, the easiest such "character" to code in

---

[*] Again, as with the CARDIAC version, with four numbers to add, you only need to enter in 3 as your first input— but that might be cumbersome to explain to the user, so entering in 4 works just as well.

will be any negative number; momentarily, we'll write a program around that terminating rule. Of course, this will mean that any negative numbers on the input card will halt the machine— so negative numbers won't be able to factor into a sum.

CARDIAC Program No. 2-5: Add *N* Numbers (with Automatic Stop)

ADDRESS	CONTENTS	MNEMONIC	OPERAND	COMMENTS
17	091	INP	91	Reads the first number into cell 91.
18	191	CLA	91	Clears the accumulator, replacing its contents with the value in cell 91.
19	092	INP	92	Reads the second number into cell 92.
20	292	ADD	92	Adds the value present in cell 92 with the contents of the accumulator.
21	691	STO	91	Stores the contents of the accumulator back into cell 91.
22	131	CLA	31	Clears the accumulator, replacing its contents with the value in cell 31 (which is zero).
23	092	INP	92	Reads the next number into cell 92.
24	192	CLA	92	Clears the accumulator, replacing its contents with the value in cell 92.
25	329	TAC	29	Tests the contents of the accumulator: if positive or zero, then simply increment the program counter; if negative, then jump to cell 29.
26	291	ADD	91	Adds the value present in the cell 91 with the contents of the accumulator.
27	691	STO	91	Stores the contents of the accumulator back into cell 91.
28	822	JMP	22	Unconditional jump to cell 22.
29	591	OUT	91	Output the contents of cell 91.
30	900	HRS	00	Halt and reset.
31	000	DAT		Data, used to reset the accumulator to zero.

To help make this program clearer to understand, examine the flowchart on the next page.

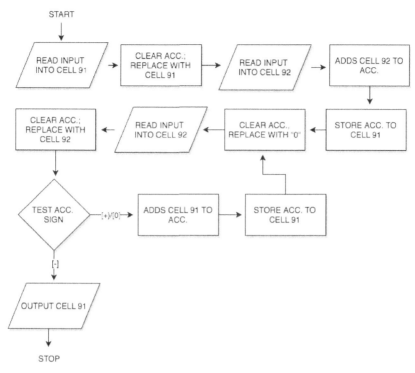

Fig. No. 8. Flowchart of CARDIAC Program No. 2-5.

Of course, an equivalent program can be made for the LMC; in fact, thanks to the several conditional jump options, we can do one better than the CARDIAC, using virtually the same amount of memory: instead of the terminating "character" being any negative number, let's make it simply zero— the only number that wouldn't affect the sum regardless. That way, negative numbers can factor into the final, calculated sum as well as positive numbers.

LMC Program No. 2-5: Add *N* Numbers (with Automatic Stop)

MAILBOX	CONTENTS	MNEMONIC	OPERAND	COMMENTS
17	901	INP		Reads the first number to be added into the calculator.
18	391	STO	91	Stores the contents of the calculator into mailbox 91.
19	901	INP		Reads the second number to be added into the calculator.
20	392	STO	92	Stores the contents of the calculator into mailbox 92.
21	191	ADD	91	Adds the value present in mailbox 91 to the value already present in the calculator.
22	391	STO	91	Stores the contents of the calculator into mailbox 91.

23	532	LDA	32	Clears the calculator; inserts value present in mailbox 32 into the calculator.
24	901	INP		Reads the next number to be added into the calculator.
25	392	STO	92	Stores the contents of the calculator into mailbox 92.
26	730	BRZ	30	Tests the contents of the calculator: if zero, jump to mailbox 30; if positive or negative, simply go to the next mailbox.
27	191	ADD	91	Adds the value present in mailbox 91 to the value already present in the calculator.
28	391	STO	91	Stores the contents of the calculator into mailbox 91.
29	623	BRA	23	Unconditional jump to mailbox 23.
30	591	LDA	91	Clears the calculator; inserts value present in mailbox 91 into the calculator.
31	902	OUT		Copies the value contained in the calculator to the outbox.
32	000	COB		Halt the machine. Also, the "000" is used as data elsewhere to reset the calculator to zero.

Note that contents of mailbox 32, which are 000, most clearly illustrates the von Neumann architecture's claim to fame: that instructions and data are treated the same way, differing only according to context.

A functional calculator should not only be able to add— it should be able to multiply as well. We will restrict our algorithm to multiplying only two numbers: the multiplier and the multiplicand. The multiplicand, after the programs subtracts it by one, will serve as a counter, keeping track of the number of times (still) necessary to add the multiplier to itself; when the counter expires (i.e., turns negative), the program will output the product and halt the machine.

CARDIAC Program No. 2-6: Multiply Two Numbers

ADDRESS	CONTENTS	MNEMONIC	OPERAND	COMMENTS
34	693	STO	93	Stores the value present in the accumulator (zero) in cell 93 for future use.
35	091	INP	91	Reads the multiplier into cell 91.
36	092	INP	92	Reads the multiplicand into cell 92.
37	192	CLA	92	Clears the accumulator; stores the multiplicand into the accumulator.
38	700	SUB	00	Subtracts the value in cell 00 (which is 001) from the value present in the accumulator.
39	692	STO	92	Stores the value present in the accumulator into cell 92.

40	193	CLA	93	Clears the accumulator; stores the value in cell 93 into the accumulator.
41	291	ADD	91	Adds the value present in cell 91 to the contents of the accumulator.
42	693	STO	93	Stores the value present in the accumulator into cell 93.
43	192	CLA	92	Clears the accumulator; stores the value in cell 92 into the accumulator.
44	700	SUB	00	Subtracts the value in cell 00 (which is 001) from the value present in the accumulator.
45	692	STO	92	Stores the value present in the accumulator into cell 92.
46	348	TAC	48	Tests the contents of the accumulator: if positive or zero, then simply increment the program counter; if negative, then jump to cell 48.
47	840	JMP	40	Unconditional jump to cell 40.
48	593	OUT	93	Output the contents of cell 93.
49	900	HRS	00	Halt and reset.

Here is the LMC version of the program:

LMC Program No. 2-6: Multiply Two Numbers

MAILBOX	CONTENTS	MNEMONIC	OPERAND	COMMENTS
34	393	STO	93	Stores the value present in the calculator (zero) for future use.
35	901	INP		Reads the multiplier into the calculator.
36	391	STO	91	Stores the contents of the calculator into mailbox 91.
37	901	INP		Reads the multiplicand into the calculator.
38	392	STO	92	Stores the contents of the calculator into mailbox 92.
39	593	LDA	93	Clears the calculator; inserts the value present in mailbox 93 into the calculator.
40	191	ADD	91	Adds the value present in mailbox 91 to the value already present in the calculator.
41	393	STO	93	Stores the contents of the calculator into mailbox 93.
42	592	LDA	92	Clears the calculator; inserts the value present in mailbox 92 into the calculator.
43	250	SUB	50	Subtracts the value present in mailbox 50 from the value already present in the calculator.
44	392	STO	92	Stores the contents of the calculator into mailbox 92.
45	747	BRZ	47	Tests the contents of the calculator: if zero, jump to mailbox 47; if positive or negative, simply go to the next mailbox.
46	639	BRA	39	Unconditional jump to mailbox 39.

ADDRESS	CONTENTS	MNEMONIC	OPERAND	COMMENTS
47	593	LDA	93	Clears the calculator; inserts the value present in mailbox 93 into the calculator.
48	902	OUT		Copies the value contained in the calculator to the outbox.
49	000	COB		Halt the machine.
50	001	DAT		Data, used to decrement the multiplicand.

So now our calculator can add (and sort of subtract) and multiply; of course, coding in division comes next.

Once a dividend and a divisor are entered on the input card, the quotient will be found by counting the number of times the divisor can be subtracted from the dividend before encountering a difference of zero— or less than zero. When the program produces the output, the quotient is relayed as a mixed number: an integer along with a proper fraction. The integer is the first number output, the numerator of the proper fraction the second number, and the denominator of the proper fraction the third (the proper fraction will not be reduced, however). For instance, enter 16 and 5— meaning you wish to divide 16 by 5— and the output of the program will be 3, 1, and 5, meaning the quotient is 3⅕. Here is the CARDIAC program:

CARDIAC Program No. 2-7: Divide Two Numbers

ADDRESS	CONTENTS	MNEMONIC	OPERAND	COMMENTS
51	692	STO	92	Stores the value present in the accumulator (zero) in cell 92 for future use.
52	090	INP	90	Reads the dividend into cell 90.
53	091	INP	91	Reads the divisor into cell 91.
54	192	CLA	92	Clears the accumulator; sets the value in the accumulator to zero.
55	200	ADD	00	Adds the value present in cell 00 (which is 001) with the contents of the accumulator.
56	692	STO	92	Stores the value present in the accumulator into cell 92.
57	190	CLA	90	Clears the accumulator; stores the value in cell 90 into the accumulator.
58	791	SUB	91	Subtracts the value in cell 91 from the value present in the accumulator.
59	690	STO	90	Stores the value present in the accumulator into cell 90.
60	362	TAC	62	Tests the contents of the accumulator: if positive or zero, then simply increment the program counter; if negative, then jump to cell 62.
61	854	JMP	54	Unconditional jump to cell 54.
62	192	CLA	92	Clears the accumulator; stores the value in cell 92 into the accumulator.

MAILBOX	CONTENTS	MNEMONIC	OPERAND	COMMENTS
63	700	SUB	00	Subtracts the value in cell 00 (which is 001) from the value present in the accumulator.
64	692	STO	92	Stores the value present in the accumulator into cell 92.
65	592	OUT	92	Output the contents of cell 92 (the integer piece of the quotient).
66	191	CLA	91	Clears the accumulator; stores the value in cell 91 into the accumulator.
67	290	ADD	90	Adds the value present in cell 90 with the contents of the accumulator.
68	693	STO	93	Stores the value present in the accumulator into cell 93.
69	593	OUT	93	Output the contents of cell 93 (the numerator of the proper fraction of the quotient).
70	591	OUT	91	Output the contents of cell 91 (the denominator of the proper fraction of the quotient).
71	900	HRS	00	Halt and reset.

The LMC version of the division algorithm, which unsurprisingly necessitates several alternations as compared to the CARDIAC's, looks like the following.

LMC Program No. 2-7: Divide Two Numbers

MAILBOX	CONTENTS	MNEMONIC	OPERAND	COMMENTS
51	392	STO	92	Stores the value present in the calculator (zero) for future use.
52	901	INP		Reads the dividend into the calculator.
53	390	STO	90	Stores the dividend into mailbox 90.
54	901	INP		Reads the divisor into the calculator.
55	391	STO	91	Stores the divisor into mailbox 91.
56	592	LDA	92	Clears the calculator; inserts the value present in mailbox 92 into the calculator.
57	172	ADD	72	Adds the value present in mailbox 72 (001) to the value already present in the calculator.
58	392	STO	92	Stores the contents of the calculator into mailbox 92.
59	590	LDA	90	Clears the calculator; inserts the value present in mailbox 90 into the calculator.
60	291	SUB	91	Subtracts the value present in mailbox 91 from the value already present in the calculator.
61	390	STO	90	Stores the contents of the calculator into mailbox 90.

62	856	BRP	56	Tests the contents of the calculator: if positive or zero, jump to mailbox 56; if negative, simply go to the next mailbox.
63	592	LDA	92	Clears the calculator; inserts the value present in mailbox 92 into the calculator.
64	272	SUB	72	Subtracts the value present in mailbox 72 from the value already present in the calculator.
65	902	OUT		Copies the value contained in the calculator to the outbox (the integer piece of the quotient).
66	591	LDA	91	Clears the calculator; inserts the value present in mailbox 91 into the calculator.
67	190	ADD	90	Adds the value present in mailbox 90 to the value already present in the calculator.
68	902	OUT		Copies the value contained in the calculator to the outbox (the numerator of the proper fraction of the quotient).
69	591	LDA	91	Clears the calculator; inserts the value present in mailbox 91 into the calculator.
70	902	OUT		Copies the value contained in the calculator to the outbox (the denominator of the proper fraction of the quotient).
71	000	COB		Halt the machine.
72	001	DAT		Data, used to increment integer portion of quotient.

If you load the N-number addition program, the multiplication program, and the division program into the CARDIAC or LMC *all at once*, you have a three-operation calculator: after entering some integers on the input card, to perform addition on them, set the program counter to 17 and then run the program; to perform multiplication, set the program counter to 34 and then run the program; or to perform division, set the program counter to 51 and then run the program.

You might even consider adding exponentiation to the calculator program as well (perhaps starting the code at memory cell 75— the other three arithmetic operations haven't left us much space in memory). The code for exponentiation, as well as for other mathematical functions of interest, is left as a challenge to the reader.

shifting

Let's put aside the LMC momentarily as we focus on the CARDIAC's shift instruction.

Shift is the only instruction with an operand that doesn't reference an address. Instead, the shift instruction (opcode - 4; mnemonic - SFT) literally shifts the value present in the accumulator x places left (where x = first digit of operand) and y places right (where y = second digit of operand). So, for example, if the value in the accumulator is 0879 and the instruction is

423   (SFT 23)

then the following occurs:

Original value in accumulator: 0879
First, shift 2 places to the left: 7900
Then, shift 3 places to the right: 0007

So, after the shift instruction, the accumulator shows 0007. Notice that with each shift, zeros replace any "empty" space, such that the accumulator always displays four digits. In fact, no matter what digits reside in the accumulator, the accumulator can always be wiped clean (set to 0000) with either of these two machine language shift instructions: 404 or 440.[*]

The CARDIAC's shift instruction can be used to (1) clear the accumulator, (2) multiply the value in the accumulator by a power of ten (perhaps losing leftmost digits in the process), or (3) Divide the value in the accumulator by a power of ten (losing the decimal portion of the number). When introducing the shift instruction, the CARDIAC manual details a program to reverse the order of the digits of a three-digit number.

The LMC, though, has no shift instruction. Let's see if we can write an LMC program to shift a number.

The rather forbidding LMC program listed below right shifts a number (up to four digits in length) zero, one, two, three, or four digits. The program accepts two inputs: first, the number you wish to shift; and second, the number of digits you want to right shift by.

LMC Program No. 2-8: Right Shift

MAILBOX	CONTENTS	MNEMONIC	OPERAND	COMMENTS
20	393	STO	93	Stores the value present in the calculator (zero) for future use.
21	901	INP		Reads the number-to-shift into the calculator.
22	391	STO	91	Stores the number-to-shift into mailbox 91.
23	901	INP		Reads the right-shift-number into the calculator.
24	392	STO	92	Stores the right-shift-number into mailbox 92.

---

[*] These also work, but are redundant (because the accumulator permits only four digits): 405, 406, 407, 408, 409; 415, 416, 417, 418, 419; and so on, as well as any three-digit number greater than 440.

*	*	*	*	
25	735	BRZ	35	If the right-shift-number is equal to zero (as shown in the calculator), then jump to mailbox 35; if not, go to the next mailbox.
*	*	*	*	
26	592	LDA	92	Load the right-shift-number into the calculator.
27	281	SUB	81	Subtract the value of 1 (in mailbox 81) from the right-shift-number.
28	738	BRZ	38	If the calculator now shows zero, jump to mailbox 38 (meaning the right-shift-number equals one).
*	*	*	*	
29	281	SUB	81	Subtract the value of 1 (in mailbox 81) a second time from the number in the calculator.
30	741	BRZ	41	If the calculator now shows zero, jump to mailbox 41 (meaning the right-shift-number equals two).
*	*	*	*	
31	281	SUB	81	Subtract the value of 1 (in mailbox 81) a third time from the number in the calculator.
32	744	BRZ	44	If the calculator now shows zero, jump to mailbox 44 (meaning the right-shift-number equals three).
33	281	SUB	81	Subtract the value of 1 (in mailbox 81) a fourth time from the number in the calculator.
34	747	BRZ	47	If the calculator now shows zero, jump to mailbox 47 (meaning the right-shift-number equals four).
35	591	LDA	91	Brings up the number-to-shift, untouched, into the calculator.
36	902	OUT		Outputs the untouched number-to-shift (since there was no shift).
37	000	COB		Halt the machine.
*	*	*	*	
38	582	LDA	82	Loads a divisor of 10 (from mailbox 82) into the calculator.
39	390	STO	90	Store this divisor into mailbox 90 for future use.
40	650	BRA	50	Jump unconditionally to the main program, starting in mailbox 50.
*	*	*	*	
41	583	LDA	83	Loads a divisor of 100 (from mailbox 83) into the calculator.
42	390	STO	90	Store this divisor into mailbox 90 for future use.
43	650	BRA	50	Jump unconditionally to the main program, starting in mailbox 50.
*	*	*	*	
44	584	LDA	84	Loads a divisor of 999 (from mailbox 84) into the calculator; notice the divisor should be 1000 but can only be 999 because we're restricted to three-digit numbers in memory.

45	390	STO	90	Store this divisor into mailbox 90 for future use.
46	650	BRA	50	Jump unconditionally to the main program, starting in mailbox 50.
*	*	*	*	
47	580	LDA	80	With a right-shift of four, the output is automatically zero; therefore, load zero (stored in mailbox 80) into the calculator.
48	902	OUT		Copies the value contained in the calculator (which is zero) to the outbox.
49	000	COB		Halt the machine.

MAIN PROGRAM

50	593	LDA	93	Load the value stored in mailbox 93 into the calculator (this mailbox will house the counter).
51	181	ADD	81	Add the number one (stored in mailbox 81) to the calculator's value.
52	393	STO	93	Store the value in the calculator back into mailbox 93.
53	591	LDA	91	Loads the value stored in mailbox 91 (initially, the number-to-shift) into the calculator.
54	290	SUB	90	Subtract the divisor by the value in the calculator.
55	391	STO	91	Store the value in the calculator back into mailbox 91.
56	850	BRP	50	If the value in the calculator is either positive or zero, jump to mailbox 50; if the value is negative, go to the next mailbox.
57	593	LDA	93	Load the counter value (in mailbox 93) into the calculator.
58	281	SUB	81	Subtract the counter by one (the value in mailbox 81); this relays the correct final answer to the right shift.
59	902	OUT		Copies the value contained in the calculator to the outbox.
60	000	COB		Halt the machine.
*	*	*	*	
80	000	DAT		Data.
81	001	DAT		Data, for adding/subtracting counters.
82	010	DAT		Data, for a divisor signaling a right shift of 1.
83	100	DAT		Data, for a divisor signaling a right shift of 2.
84	999	DAT		Data, for a divisor signaling a right shift of 3.

An LMC program that left shifts is marginally easier to write than one that right shifts, since, in effect, you will have to multiply rather than divide.

There are a number of ways to go about coding a right-shift LMC program; the exercise is left as a challenge to the reader.

subroutines

In the book *Liber Abaci*, the mathematician Fibonacci (also known as Leonardo of Pisa) presents a simple sequence modeling the growth of a population of rabbits. Suppose that the gestation period of a rabbit is only one month. We start off with a single pair, and at the end of the month there is still only a single pair— but they have mated and will produce offspring at the end of the second month, making two pairs: the original set and a new set of rabbits. Another month passes, and another pair is produced, for a total of three pairs of rabbits. This continues indefinitely.

The sequence— later called the Fibonacci sequence— with its humble beginnings quickly became famous because of its simplicity: it is an exercise in mathematical recursion, with each term being the sum of the previous two (save for the first two terms, which are 1 and 1). Thus, the sequence is 1, 1, 2, 3, 5, 8, 13, 21, and so on.

Writing a program to output the Fibonacci sequence requires coding several repetitive tasks. We might therefore consider writing a subroutine for the program to call every time a particular (repetitive) task needs to be executed. Consider the CARDIAC program below:

CARDIAC Program No. 2-8: Fibonacci Sequence

ADDRESS	CONTENTS	MNEMONIC	OPERAND	COMMENTS
17	100	CLA	00	Clears the accumulator; stores the value in cell 00 (which is 001) into the accumulator.
18	690	STO	90	Stores the value present in the accumulator into cell 90.
19	691	STO	91	Stores the value present in the accumulator into cell 91.
20	590	OUT	90	Output the contents of cell 90.
21	591	OUT	91	Output the contents of cell 91.

	MAIN PROGRAM			
22	190	CLA	90	Clears the accumulator; stores the value of cell 90 into the accumulator.
23	291	ADD	91	Adds the value present in cell 91 with the contents of the accumulator.
24	692	STO	92	Stores the value present in the accumulator into cell 92.
25	592	OUT	92	Output the contents of cell 92.
26	895	JMP	95	Unconditional jump to cell 95 (the beginning of the subroutine).
27	822	JMP	34	Unconditional jump to cell 22 (the beginning of the main program).
	SUBROUTINE			
95	191	CLA	91	Clears the accumulator; stores the value of cell 91 into the accumulator.
96	690	STO	90	Stores the value present in the accumulator into cell 90.
97	192	CLA	92	Clears the accumulator; stores the value of cell 92 into the accumulator.
98	691	STO	91	Stores the value present in the accumulator into cell 91.
99	8--	JMP	--	Performs an unconditional jump to the return address— which, when the CARDIAC executes this instruction, should be cell 27.

The program takes advantage of the CARDIAC's "save state"* feature, permitting it to return from the subroutine call, by abutting all of the subroutine's code next to memory cell 99. There is an obvious limitation of this feature: since the return address is overwritten with each successive jump instruction, only one "save state" is retrievable at any given time.

Although the above program could easily be translated to the LMC, the LMC has no "save state" feature—and thus no way to directly call and return from a subroutine.[†]

---

* In terms of standard computer science lexicon, I'm using this term fairly loosely here—which is why I keep enclosing the term in quotes—because it compactly describes what I mean. Note that a "save state" usually refers to some event that, because it has been interrupted, is placed on top of a memory stack for later reference. Technically, the CARDIAC's unconditional jump really is a call/return instruction: the call instruction jumps to a new address but not before saving the prior address (+ 1), then pushes this saved address to the top of the stack. The return instruction finally pops this saved address off the stack.

† But some nonstandard LMC instruction sets have added call and return instructions. For example, the Little Man Computer on the Raspberry Pi uses opcode 4 to jump to a subroutine and numeric code 903 to return from the subroutine. Another popular extension of the LMC paradigm is called the Son-of-LMC, which also permits subroutine calls/returns.

bootstrapping

There's been something unrealistic about how we've been presenting and running CARDIAC programs thus far. Stenciling in— or typing in, using an emulator— the program's instructions (the data words) into their corresponding memory cells ignores a central notion about modern computers: programs must be loaded into memory by other programs *before* they can be run. We can load in a program via— what else?— the input.

And this is where the brilliance of the word "001" hardwired into permanent memory (ROM) in memory cell 000 becomes apparent. It can be used for much more than just incrementing or decrementing a counter by one unit. Courtesy of a two-line bootstrapping routine, along with alternating address/data word instructions, we can load an entire program into memory automatically. The bootstrapping routine is simple: inputs of 002, then 800, and then, at the end, 8xx, where xx is the starting address of the program.

Here's how it works: 001 (hardwired into memory cell 000) reads the first input into memory cell 001. That first input, 002, reads the second input into memory cell 002. The second input is 800— an unconditional jump back to memory cell 000, which contains 001. And we're back where we started, which doesn't seem like progress. Except the CARDIAC is now ready to load the next instruction into memory. That next instruction must have the *address* of the program's first instruction, while the following input must contain the *data word* to insert into the aforementioned address.

We will call the entire input (including 002 and 800), which has loading instructions and data, an "input stack." The input stack contains the following data cards, in the order listed below.

    002
    800
    First Address
    First Data Word
    Second Address
    Second Data Word
    Third Address
    Third Data Word
    and so on...until:
    8xx, where xx = First Address

Perhaps a fishing metaphor will help make the process clearer. The input stack, swimming in an ocean of data, is reeled in by the "hook" at dangling at the end of the CARDIAC: the data word 001 in ROM. Without that hook, no input stack— and, therefore, no program— can be retrieved from the vast ocean of data.

Let's try an example. Glance again at this section's first CARDIAC program:

CARDIAC Program No. 2-1: Add Two Numbers

ADDRESS	CONTENTS	MNEMONIC	OPCODE	COMMENTS
17	090	INP	90	Read the input card into cell 90.
18	091	INP	91	Read the input card into cell 91.
19	190	CLA	90	Clear the accumulator, setting its contents to be the data in cell 90.
20	291	ADD	91	Add contents of cell 91 to the accumulator.
21	692	STO	92	Store the contents of the accumulator into cell 92.
22	592	OUT	92	Output the contents of cell 92.
23	900	HRS	00	Halt and reset.

To transfer this program into memory through the input— which, when the CARDIAC "boots," has only the data 001 in memory cell 000— we load the following input stack:[*]

002
800
017
090
018
091
019
190
020
291
021
692
022
592
023
900
817

Note that since our program adds two numbers retrieved from input, we need two more inputs after the 817: the first and second numbers we wish to add.[†]

---

[*] We could also refer to the input stack as a "card deck," since inputs cards were fed, one at a time, into the antediluvian computers that the CARDIAC models— and a stack of cards can be considered a deck.

[†] It may seem like I've contradicted myself, since several paragraphs ago I wrote, "...and then, *at the end*, 8xx, where xx is the starting address of the program." Except that my "at the end" phrase refers to the bootstrapping routine itself, not to the entire input stack, which, for this particular adding-two-numbers program, requires two additional inputs: the two numbers you wish to add.

Unless you're a purist or a masochist, it's not necessary to use a bootstrapping routine every time (or anytime) you wish to run a CARDIAC program (although you could); rather, we can continue to copy the data words directly into their corresponding memory cells, albeit with the knowledge that real computers don't quite do it that way.

Before moving on, let's turn to the LMC: does it have a commensurate bootstrapping routine? Since the LMC has no ROM, only one hundred memory cells (mailboxes) of RAM, the machine has no "hook" to reel in an input stack; hence, for an unmodified LMC, we'll always have to (unrealistically but conveniently) encode programs directly into their corresponding memory cells. Thus, any possible LMC bootloader would have to perforce treat some RAM mailboxes as "ROM," effectively changing how to program on the LMC. (Such a move would, of course, be non-canonical.) For instance, one attempt to write an LMC bootloader might set aside bootstrap code in mailboxes 80 to 99, labeling these locations "ROM," and loading programs one instruction at a time, starting at mailbox 00, through the inbox; in addition, to make it all work, the LMC's program counter needs always to be set to start at 80.[*]

There are surely many other possible LMC bootloader routines, but they all by necessity violate one key LMC assumption: that all mailboxes are RAM.

assembly language programming

Managing the addresses (or mailboxes) of data, counters, and the like can quickly get complicated; shift a block of code in memory, and references to other memory cells (e.g., jump instructions) may need to change.[†]

With assembly language, though, we can use "labels," which allows the computer to set aside and manage memory locations out of sight. The bookkeeping is automatically taken care of. We also can construct variables to hold data. All of the mnemonics of the instruction set function the same way as before, along with one extra one: DAT. The DAT instruction is dual purpose: (1) It can declare a variable by setting aside memory for it, and (2) It can set a variable equal to a particular number.

Consider the following simple CARDIAC program. A single input number is required. The program then decrements the input by one unit at a time until reaching -1, at which point the -1 is output to the user and the program terminates.

---

[*] http://teaching.idallen.com/dat2343/10f/notes/364_LMC_bootstrap.html

[†] The CARDIAC manual mentions CARDIAC assembly language, albeit briefly and only as contrast to the high-level FORTRAN language (not to spur you on to write CARDIAC code in assembly), even offering a five-line CARDIAC assembly program to work through. Note: A high-level language similar to FORTRAN written in CARDIAC assembly, which is a product of CARDIAC machine language, is technically possible but in practice infeasible because of the CARDIAC's memory limitations (the same goes for the LMC).

CARDIAC Program No. 2-9: Decrement Counter

ADDRESS	CONTENTS	MNEMONIC	OPERAND	COMMENTS
17	090	INP	90	Value is input, and assigned to cell 90.
18	190	CLA	90	Accumulator is cleared, and loaded with value in cell 90.
19	700	SUB	00	The value in cell 00 (which is 001) is subtracted from the value in the accumulator.
20	690	STO	90	The value in the accumulator is stored back into cell 90.
21	323	TAC	23	If the accumulator shows a positive number or zero, go on to the next cell; if the accumulator is negative, jump to cell 23.
22	818	JMP	18	An unconditional jump to cell 18.
23	590	OUT	90	Output the value stored in cell 90.
24	900	HRS	00	Halt the machine.

This machine language program can be shed of its reliance on specific address assignments by writing it in assembly code instead:

CARDIAC Program No. 2-10: Decrement Counter, Again

LABEL	MNEMONIC	OPERAND	COMMENTS
Start	INP	counter	Asks the user for input; stores this input into the variable "counter."
loop	CLA	counter	Clears the accumulator's contents, replacing them with the value of the variable "counter."
	SUB	constant	Adds the contents of the variable "constant" (which is 1) to the contents in the accumulator.
	STO	counter	Stores the contents of the accumulator back into the variable "counter."
	TAC	output	If the accumulator is positive or zero, goes on to the next instruction; if it is negative, jump to the "output" label.
	JMP	loop	Unconditional jump to the "loop" label.
output	OUT	counter	Prints the value of the variable "counter."
	HRS		End the program
constant	DAT	1	Assign the variable "constant" to an initial value of 1.

Once untethered to the explicit address assignments of the operands, notice how much cleaner, organized, and easier to follow the assembly code is than the machine code. The code and variable labels organize the program, while strict use of mnemonics facilitates readability.

Like the CARDIAC, the LMC can also receive the assembly language treatment. Here is the equivalent LMC assembly program:

LMC Program No. 2-9: Decrement Counter

LABEL	MNEMONIC	OPERAND	COMMENTS
start	INP		Asks the user for the first input; copies the input into the calculator.
	STA	counter	Stores the contents of the calculator into the variable "counter."
loop	SUB	constant	Subtracts the contents of the calculator by the value of the variable "constant" (which equals 1).
	STA	counter	Stores the contents of the calculator back into the variable "counter."
	BRP	loop	If the calculator shows contents that are positive or zero, jump to the "loop" label; otherwise, move on to the next instruction.
	OUT		Copy the value contained in the calculator to the outbox.
	COB		End the program.
constant	DAT	1	Assigns the variable "constant" an initial value of 1.

Assembly's simpler structure really pays off when writing more complicated programs. For instance, here's a CARDIAC program for squaring a number input by the user; as you work through it, keep in mind how much more intuitive it reads than the corresponding machine language version:

CARDIAC Program No. 2-11: Square a Number

LABEL	MNEMONIC	OPERAND	COMMENTS
start	INP	base	Asks the user for the first input; stores the input into the variable "base."
	CLA	base	Clears the accumulator's contents, replacing them with the value of the variable "base."
	SUB	one	Subtracts the contents of the accumulator by the value of the variable "one" (which equals 1).
	STO	multiplier	Stores the contents of the accumulator into the variable "multiplier."
	CLA	zero	Clears the accumulator's contents, replacing them with the value of the variable "zero" (which equals 0).
loop	CLA	result	Clears the accumulator's contents, replacing them with the value of the variable "result."
	ADD	base	Adds the contents of the variable "base" to the contents in the accumulator.
	STO	result	Stores the contents of the accumulator into the variable "result."
	CLA	multiplier	Clears the accumulator's contents, replacing them with the value of the variable "multiplier."
	SUB	one	Subtracts the contents of the accumulator by the value of the variable "one" (which equals 1).
	STO	multiplier	Stores the contents of the accumulator into the variable "multiplier."
	TAC	end	If the accumulator is positive or zero, goes on to the next instruction; if it is negative, jump to the "output" label.
	JMP	loop	Unconditional jump to the "loop" label.
end	OUT	result	Prints the value of the variable "result."
	HRS		End the program.

one	DAT	1	Assign the variable "one" to an initial value of 1.
zero	DAT	0	Assign the variable "zero" to an initial value of 1.
result	DAT		Define address for the variable "result."

Here is the corresponding LMC assembly program to square a number:

LMC Program No. 2-10: Square a Number

LABEL	MNEMONIC	OPERAND	COMMENTS
start	INP		Asks the user for the first input; copies the input into the calculator.
	STA	base	Stores the contents of the calculator into the variable "base."
	SUB	one	Subtracts the contents of the calculator by the value of the variable "one" (which is equal to 1).
	STA	multiplier	Stores the contents of the calculator into the variable "multiplier."
	LDA	zero	Loads the value of the variable "zero" (which is equal to zero) into the calculator.
loop	LDA	result	Loads the value of the variable "result" into the calculator.
	ADD	base	Adds the contents of the variable "base" to the contents in the calculator.
	STA	result	Stores the contents of the calculator into the variable "result."
	LDA	multiplier	Loads the value of the variable "multiplier" into the calculator.
	SUB	one	Subtracts the contents of the calculator by the value of the variable "one" (which is equal to 1).
	STA	multiplier	Stores the contents of the calculator into the variable "multiplier."
	BRP	loop	If the calculator shows contents that are positive or zero, jump to the "loop" label; otherwise, move on to the next instruction.
end	LDA	result	Loads the value of the variable "result" into the calculator.
	OUT		Copy the value in contained in the calculator to the outbox.
	COB		End the program
one	DAT	1	Assigns the variable "one" an initial value of 1.
zero	DAT	0	Assigns the variable "zero" an initial value of 0.
result	DAT		Define address for the variable "result."

Want more of a challenge than simply squaring a number? Write a program to accept two inputs for exponentiation: a base and a power. It's not as straightforward as it may initially seem.

recursion

In mathematics, a recursive function is a function which calls itself: the most recent output from the function becomes the next input for it. In computer science, the definition of recursion is similar: a procedure or subroutine which calls itself. We've already seen some programs that, at least informally, already do so. Let's see if we can exploit recursion in a CARDIAC/LMC program.

Consider an algorithm to count down the numbers from *N* to 0. The associated recursive function, written in high-level language pseudocode, might look like this:

```
function countdown(int N)
   if N >= 0:
     countdown(N-1)
     output N;
```

The function countdown, which calls itself, requires a parameter when called: the initial value of the countdown.

For instance, if countdown(10) is called, the output is as follows:

```
10
9
8
7
6
5
4
3
2
1
0
```

And since the function only outputs numbers when the count is at least zero, the recursion— and the function call— ends.

What follows is a CARDIAC program that does the same thing, in a similar way. Notice that the program employs an assembly syntax we haven't yet encountered: a number as an operand.

CARDIAC Program No. 2-12: Countdown

LABEL	MNEMONIC	OPERAND	COMMENTS
start	INP	max	Asks the user for the maximum number— i.e., the number with which to begin the countdown; it stores this number into the variable "max."
	CLA	max	Clears the accumulator's contents, replacing them with the value of the variable "max."
	ADD	1	Adds 1 to the value in the accumulator.
countdown	SUB	1	Subtracts 1 from the value in the accumulator.
	TAC	end	If the accumulator is positive or zero, goes on to the next instruction; if it is negative, jump to the "end" label.
	STO	result	Stores the contents of the accumulator into the variable "result."
	OUT	result	Prints the value of the variable "result."

	JMP	countdown	Unconditional jump to the "countdown" label.
end	HRS		End the program.
max	DAT		Define address for the variable "max."

The LMC version looks like this:

LMC Program No. 2-11: Countdown

LABEL	MNEMONIC	OPERAND	COMMENTS
start	INP		Asks the user for the maximum number— i.e., the place to begin the countdown; it places the input into the calculator.
	ADD	1	Adds 1 to the value in the calculator.
countdown	SUB	1	Subtracts 1 from the value in the calculator.
	OUT		Copy the value contained in the calculator to the outbox.
	BRZ	end	If the calculator shows zero, then jump to the "end" label; otherwise, move on to the next instruction.
	BRA	countdown	Unconditional jump to the "countdown" label.
end	COB		End the program.

reference: even more advanced machine language routines

For a taste of even more machine- and assembly-language routines possible with the CARDIAC (and by extension, to some extent, with the LMC), look no further than an article[*] by Brian L. Stuart, a computer science professor at Drexel University and former computer engineer.[†] Despite its limitations, the CARDIAC is still receptive to some rather sophisticated programming techniques; Stuart's primer demonstrates stacks, multiple subroutines courtesy of "subroutine linkage," recursion, a sorting algorithm that reverses the order of an inputted stack, as well as two programs that tax the hardware (or card-ware) to its core: solving the mathematical Tower of Hanoi puzzle and generating Pythagorean triples.

an encore: back to binary

Recall our earlier discussion of base-to-base conversions. How about an encore?

To convert from base 10 (decimal) to base 2 (binary) requires repeated division, examining the remainders at every turn; those remainders form the string of digits of the number in binary— but reverse order.

---

[*] https://www.cs.drexel.edu/~bls96/museum/cardiac.html

[†] https://www.cs.drexel.edu/~bls96/

This convert-to-binary program serves as a nice review of a number of programming ideas we've discussed so far. Recursion informally comes into play starting from the second division: the dividend is equal to the prior quotient; the output turns into the next input. Our code below will apportion the arithmetical processes into subroutines.

CARDIAC Program No. 2-13: Decimal to Binary Converter

LABEL	MNEMONIC	OPERAND	COMMENTS
start	STO	counter	Set the value of the "counter" variable to zero.
	INP	dividend	Stores the value of a user input into the variable "dividend."
divide	CLA	counter	Clears the accumulator, replacing its contents with the value of the "counter" variable.
	ADD	1	Increments the "counter" variable by one unit.
	STO	counter	Stores the value in the accumulator into the "counter" variable.
	CLA	dividend	Clears the accumulator, replacing its contents with the value of the "dividend" variable.
	SUB	2	Subtracts 2 from the value in the accumulator ("dividend").
	STO	dividend	Stores the value in the accumulator into the "dividend" variable.
	TAC	reportrem	If the value of the accumulator is negative, then jump to the "reportrem" label; if not, then go to the next line.
	JMP	divide	Unconditional jump to the "divide" label.
reportrem	ADD	2	Adds 2 to the value in the accumulator.
	STO	remainder	Stores the value in the accumulator into the "remainder" variable.
	OUT	remainder	Outputs the "remainder" variable.
	CLA	counter	Clears the accumulator, replacing its contents with the value of the "counter" variable.
	SUB	1	Subtracts 1 from the value in the accumulator.
	STO	dividend	Stores the value in the accumulator into the "dividend" variable.
	SUB	1	Subtracts 1 from the value in the accumulator.
	TAC	end	If the value of the accumulator is negative, then jump to the "end" label; if not, then go to the next line.
	CLA	zero	Clears the accumulator.
	STO	counter	Stores the value in the accumulator (zero) into the "counter" variable.
	JMP	divide	Unconditional jump to the "divide" label.
end	HRS		End the program.
zero	DAT	0	Assigns the variable "zero" an initial value of 0.
counter	DAT		Define address for the variable "counter."
remainder	DAT		Define address for the variable "remainder."
dividend	DAT		Define address for the variable "dividend."

Input a number into the program—say, 84—and the output, one digit at a time, will be the base 10 number converted into base 2—in this case, **1010100** (which will output as **0, 0, 1, 0, 1, 0, 1**). How's that for an encore?

Writing an LMC decimal-to-binary converter is up to you.

section

the instructo paper computer

↳ **Run IPC Machine Language Programs by Using**
The Pink Paper Computer (see Section 5)
Instructo Emulator (see Section 8)
Original or Replica IPC Hardware
Or Your Mind

↳ **Run IPC Assembly Language Programs by Using**
Instructo Compiler (see Section 8)
Or Your Mind

The Instructo is burdened by an excessive number of moving parts and a large instruction set.* But its complexity also permits the user to create and run more sophisticated programs with fewer instructions than the CARDIAC or LMC, and in several key ways it may be the paper computer that most faithfully models a modern electronic binary computer.

the manual

Snap open up the Instructo operator's manual, and—compared to the CARDIAC manual, at any rate—you're in for a bit of an antiseptic, sterile, impersonal experience. That's not to say it's poorly written; it is most certainly not. The operator's manual is thorough and complete and goes out of its way, especially in the early going, to copiously explain every programming detail. But it lacks the style and the instructional cartoon eye candy of the CARDIAC manual.

Like Hagelbarger's manual, Matt opens his with an appeal explaining the importance of keeping up with technological advancement so as to not get left behind:

---

* Even though I have a nostalgic connection to the Instructo as well as the greatest respect for its creator, my criticism of the concept and the device itself will be unsparing.

In the world of the future most people will be computer literate. It will be important for people to know how to operate and use computers, because computers will affect all aspects of our lives. The future is yours, and this computer is a first step toward computer literacy.

Next comes an overview of the contents of the kit you've just purchased— a paper computer with all its moving slides and parts and trappings— followed by assembly instructions that vaguely recall the CARDIAC's. Paperclips, however, are needed to secure the program code to the device (each program is printed on a thin column of paper). Before a program is run, a Main Storage Unit sheet, with blanks cells for memory locations 90 to 99, must also be paper-clipped to the Main Storage Unit on the front face of the Instructo. Although the computer has a total of one hundred storage locations, memory cells 00 to 89 are designated exclusively for program storage in the Program Storage Unit (where the printed program is paper-clipped); but no storage locations are preset to ROM, akin to the CARDIAC's memory cells 00 and 99, thus precluding an innate bootstrapping routine.* The memory cells are usually referred to as "storage locations," abbreviated as SS. But the Instructo radically departs from the CARDIAC and LMC with the possible *contents* of these storage locations, since they can be more than just numbers— letters, words, or symbols are all possible.

Matt spills some ink explaining how computers work, after assuring us that the Instructo "can do almost anything a real computer can do, but it works using paper and pencil instead of electricity and integrated circuits"— which quite a claim, completely antithetical to the CARDIAC manual's approach that takes pains to emphasize that the CARDIAC isn't a real computer, but merely an aid to understanding computers.

Computers, like cars, televisions, and other "complicated devices," usually have modular designs, meaning they are made of any number of separate devices operating in tandem. If the central conceit of the LMC is that of a little man running around flipping switching and turning dials, then the central conceit of the Instructo is modularity: separate systems all working together, communicating through data busses. Interconnected busses are shown as flowchart-like arrows connecting inputs, outputs, registers, and memory together at the front of the Instructo. Some of those bus connections travel in only one direction (such as from input to main storage), while others are like two-way streets (such as from program storage to main storage), signified by the direction the arrowheads point. Pride of place is given to the path of data flow throughout the operator's manual.

---

* Like with the LMC, a bootloader that sets aside certain storage locations for ROM and permits programs to load through input might be coded, but that would require hacking the hardware (or card-ware) in a non-canonical manner.

Unlike the CARDIAC and LMC, the Instructo has two places for input: Input A and Input B. Input A is restricted to a single decimal number, while Input B uses an input strip for multiple inputs. There are two places for output as well: Output A and Output B.

The ALU (arithmetic and logic unit), rather than an accumulator as in the CARDIAC and LMC, consists of three components: Register A, Register B, and a Compare Unit. The registers can hold information from any storage location— meaning numbers, letters, words, symbols, or some combination thereof. The Compare Unit allows for less than, greater than, or equal to comparisons between numerical values contained in either of the registers and values in set storage locations. Note that either of the registers can double as an accumulator.

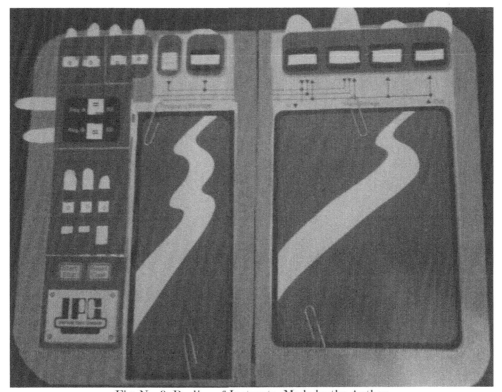

Fig. No. 9. Replica of Instructo, Made by the Author.

The final piece of the Instructo's modular puzzle is the Control Unit, housing the Index Counter, the Program Step Indicator, and the Jump Switches. The Index Counter is simply a counter on standby that can be incremented or decremented if needed. The Program Step Indicator relays to the ALU the "current" step of the program being run. And the Jump Switches, of which there are three, can assume one of two values: 0 or 1. They are versatile and can direct the computer to change course during a program, depending on the values they're showing. Jump Switches have no easy analogue[*] with the CARDIAC or LMC; perhaps they are best compared to conditional jumps resulting from the sign of the accumulator's contents. Notice also that the three Jump Switches, which allow the Instructo to relay a three-bit binary number such as 010 (decimal = 2) or 111 (decimal = 7). (There are eight possible numbers: 0 to 7, in decimal.)

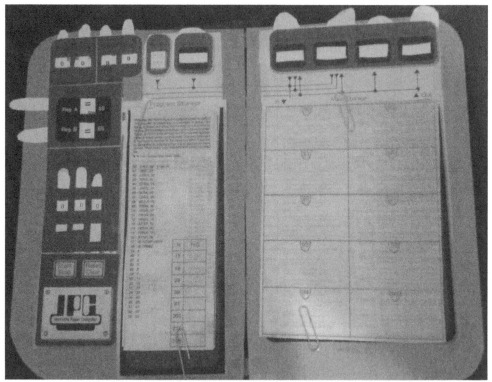

Fig. No. 10. Author-Made Instructo Replica, with a Program Loaded.

Once a program is loaded into the Program Storage Unit, the Start/Stop Switch begins the action: lots of slides to shift back and forth, and fair amount of writing and erasing and rewriting, characterize a typical program run. The Instructo bears the weight of a greater complexity than the CARDIAC, resulting in quite a bit of physical tedium when working a program through to its completion (this time *you'll* be the exhausted little man or woman). For example, early on, the operator's manual offers a guide to running a sample program. All the program does is output "HELLO, I'M A COMPUTER," but two pages of step-by-step instructions are necessary— and that's not even counting the explanatory *third* page, which details a generalized "IPC OPERATION FLOWCHART" that is required reading (since it

---

[*] And not because they're binary.

involves setting up and resetting values on the Instructo, like the Program Step Indicator and Index Counter, before each program run).

The remainder of the manual is taken up by example programs, most of which are math-related or -themed[*]— such as code to find the multiples of some number N (program #5), to calculate the average of a group of input numbers (program #7), to obtain the sum of the integers from 1 to N (program #10), to list Fibonacci numbers (program #12), and to identify prime numbers (program #14)— as well as designated outputs from those programs and suggestions for future programs.

the mnemonic codes

But it's here, with the mnemonics, that things start to get really complicated. Instead of numerical opcodes, the Instructo relies strictly on mnemonics. When running a program, there's quite a bit to juggle in your mind. For example, does the mnemonic refer to Register A or B? or Input A or B? or Output A or B? or Jump Switch A, B, or C? And then there are the more sophisticated functions you'll need to be familiar with, such as exponentiation, square roots, and digital roots, that are built in to the machine.

Instructo Instruction Set[†]

MNEMONIC(S)	DESCRIPTION
ADDA/ADDB	Sums the numbers in location SS and Register A/B, placing the result into Register A/B.
CPRA/CPRB	Determines if the number in Register A/B is >, <, or - to the number location SS; sets the result into the Register A/B Compare Unit.
DIVA/DIVB	Takes Register A/B divided by the number in location SS, placing the quotient into Register A/B and the remainder into Register B/A.
DRTA/DRTB	Finds the digital root of the number in Register A/B, placing the result into Register A/B.
DVDA/DVDB	Takes the decimal number in Register A/B and divides it by the decimal number in location SS, placing the quotient into Register A/B.
ENIA/ENIB	Reads the value of Input A/B, and sends that data to location SS.
EXPA/EXPB	Evaluates (Register A/B)^(SS). (Exponentiation.)
INDA	Sums the numbers in the Index Counter and location SS, placing the result into the Index Counter.

---

[*] Although the Instructo's operator's manual functions well as a standalone text, complete with many example programs, Matt also published a small set of supplementary books for the machine called *Programs for the Instructo*, which were libraries of programs all with a common theme, such as geometry and sports.

[†] Note that the one key change I've made in presenting the instruction set, compared to the Instructo's operator's manual, is that I have listed out all permutations of each instruction; the operator's manual instead uses a * (asterisk) to stand in for a letter (like A, B, or C) and the # (number sign) to represent either 0 or 1. For instance, PROA/PROB is shown in the operator's manual as PRO*, while the numerous Set Jump Switch instructions are written in simply as SJ*#.

INDL	Replaces the number in the Index Counter with the number in location SS.
INDS	Takes the Index Counter minus the number in location SS, placing the result into the Index Counter.
JAEQ	Conditional jump: If Register A Compare Unit shows =, jump to SS.
JAGT	Conditional jump: If Register A Compare Unit shows >, jump to SS.
JALT	Conditional jump: If Register A Compare Unit shows <, jump to SS.
JANE	Conditional jump: If Register A Compare Unit doesn't show =, jump to SS.
JANG	Conditional jump: If Register A Compare Unit doesn't show >, jump to SS.
JANL	Conditional jump: If Register A Compare Unit doesn't show <, jump to SS.
JANZ	Conditional jump: go to location SS if Register A does not equal 0.
JAZE	Conditional jump: go to location SS if Register A equals 0.
JBEQ	Conditional jump: If Register B Compare Unit shows =, jump to SS.
JBGT	Conditional jump: If Register B Compare Unit shows >, jump to SS.
JBLT	Conditional jump: If Register B Compare Unit shows <, jump to SS.
JBNE	Conditional jump: If Register B Compare Unit doesn't show =, jump to SS.
JBNG	Conditional jump: If Register B Compare Unit doesn't show >, jump to SS.
JBNL	Conditional jump: If Register B Compare Unit doesn't show <, jump to SS.
JBNZ	Conditional jump: go to location SS if Register B does not equal 0.
JBZE	Conditional jump: go to location SS if Register B equals 0.
JIBD	If there is any data in Input B, jump to location SS.
JINZ	Conditional jump: go to location SS if the Index Counter does not equal 0.
JIZE	Conditional jump: go to location SS if the Index Counter equals 0.
JJA0	If Jump Switch A is set to 0, jump to location SS.
JJA1	If Jump Switch A is set to 1, jump to location SS.
JJB0	If Jump Switch B is set to 0, jump to location SS.
JJB1	If Jump Switch B is set to 1, jump to location SS.
JJC0	If Jump Switch C is set to 0, jump to location SS.
JJC1	If Jump Switch C is set to 1, jump to location SS.
JUMP	Unconditional jump to location SS.
LDRA/LDRB	Sends the data located in storage location SS to Register A/B.
MULA/MULB	Takes Register A/B times the number in location SS, placing the result into the Register A/B.
NOOP	Increments the Program Step by one unit.
PABA	Displays a mixed fraction in Output A; specifically, the whole number part comes from Register A, the numerator comes from Register B, and the denominator comes from location SS.
PABB	Displays a mixed fraction in Output B; specifically, the whole number part comes from Register A, the numerator comes from Register B, and the denominator comes from location SS.
PBAA	Displays a mixed fraction in Output A; specifically, the whole number part comes from Register B, the numerator comes from Register A, and the denominator comes from location SS.
PBAB	Displays a mixed fraction in Output B; specifically, the whole number part comes from Register B, the numerator comes from Register A, and the denominator comes from location SS.
PROA/PROB	Prints the data in storage location SS to Output A/B.
REVA/REVB	Completely reverses the digits of the number in Register A/B.
SJA0	Sets Jump Switch A to 0.
SJA1	Sets Jump Switch A to 1.
SJB0	Sets Jump Switch B to 0.
SJB1	Sets Jump Switch B to 1.

SJC0	Sets Jump Switch C to 0.
SJC1	Sets Jump Switch C to 1.
SQTA/SQTB	Takes the square root of the number in Register A/B, placing the result back into Register A/B.
STRA/STRB	Sends the data in Register A/B to location SS.
SUBA/SUBB	Takes Register A/B minus the number in location SS, placing the result into Register A/B.
SWAP	Switch the numbers in Registers A and B.

Granted, there's a fair amount of repetition in the instruction set— which, unlike the CARDIAC/LMC's Reduced Instruction Set Computing (RISC) paradigm, more nearly resembles a Complex Instruction Set Computing (CISC) processor, like the Intel 8080— but the sheer number of permutations of mnemonics is overwhelming. And, at least to those who are accustomed to the CARDIAC/LMC paradigm, a great many mnemonics may seem at first blush unnecessary, or even redundant, while others are conspicuously absent (such as a call/return instruction for subroutines). For example, multiplication and division: Can't those just be made into subroutines in a program? And ditto exponentiation? Furthermore, what's the point of having a mnemonic expressly designed to reverse the digits in a register? (The CARDIAC manual presents a short program designed to reverse the order a three-digit number using the shift instruction; there's of course no CARDIAC instruction to reverse the digits.)

With respect to the mathematics functions especially, the answers lie with shortcutting math-related problems, since the Instructo was designed to be (mostly) used in a math classroom anyway (the majority of the programs written for the Instructo in the operator's manual are math-centric). So, for example, rather than having to write a subroutine for every time you wished to exponentiate— which is a common operation— it's pre-programmed.

Some of the mnemonics, which at first glance seem redundant, are in fact thoughtful. Mnemonics associated with the Compare Unit, for instance, compare favorably to the LMC's conditional jumps, as well as the CMP mnemonic of the paradigmatic mid-seventies Intel 8080 chip,[*] which compares the contents of the accumulator with data in a register. The Instructo's mnemonics related to displaying fractions are perhaps analogous to the Intel's DAA mnemonic, which works with decimals. Although it seems somewhat silly here, the Instructo's NOOP mnemonic, which effectively does nothing, corresponds to Intel's NOP mnemonic, which means "no operation." And, finally, later Intel chips included multiply and divide instruction codes, thus avoiding the need for subroutines, à la the Instructo.[†]

---

[*] Used in the first personal computer, 1974's primitive Altair 8800 (complete with switches and LED display, similar visually to Hagelbarger's outguessing machine from twenty years prior), which lead Bill Gates and Paul Allen— after writing Altair BASIC— to form what would later turn into Microsoft.

[†] And some other powerful chips, like the Hitachi 6309, also had multiplication and division

But the syntax of the Instructo's mnemonics still seems curious. Recall that the general format for an instruction is

Mnemonic [destination],[source]

For instance, the following instruction will store a value from the accumulator into a designated address:

STO [xx],A

But the Instructo is not a single-accumulator machine; rather, it has two registers, A and B. Now, consider the STRA/STRB mnemonics (shown in the table several pages ago). The following instruction reads the data found in Register A and sends it to an address: location 50:

STRA, 50

while the next instruction reads the data found in Register B and sends it to a different address: location 51:

STRB, 51

But the syntax seems redundant (and the comma extraneous). Wouldn't it make more sense simply for the mnemonic STR to be used instead, with the above instructions then reading this way:

STR 50,RA
STR 51,RB

The "RA" and "RB" refer, of course, to Registers A and B. (The "R" is necessary so as to avoid confusion with the term "Accumulator.") Similar issues exist with a lot of the other mnemonics— such as LDRA/LDRB, PROA/PROB, REVA/REVB, and so on— meaning the total number of mnemonics could have been reduced, although the Instructo's syntax (like much of its instruction set) is hardly unprecedented. For instance, the Motorola 6809 microchip, which powered best-selling computers like the Tandy Color Computer, had instructions like LDY (load the Y register) and ADDA, ADDB, and ADDD (sum the operand with the contents of the A, B, or D registers, respectively). Nonetheless, viewed in retrospect, reducing the number of mnemonics by using a simpler format for the instruction set repre-

---

built in to their instruction sets, so including powerful mathematical functions as instructions wouldn't have been unprecedented or unrealistic at the time of the Instructo's release.

sents a significant missed opportunity that may have made the Instructo more palatable (and saleable) to the technological neophyte.

Yet what might have been appealing to the technological neophyte— or, indeed, to the technophobe— was probably just as quickly a turn off to the technophile: the Instructo's near-complete disregard for variable types. There are, in effect, *no restrictions* as to what can be placed in the storage locations.* For example, consider program #2: "This program," the description above the code reads, "will print several important dates from the history of the United States." Except that most of the storage locations contain the names of the historical events (and places) themselves:

<pre>
         32 GOLD RUSH
         33 PHILADELPHIA
         34 INDEPENDENCE
         35 CIVIL WAR
</pre>

and so on. The Instructo's laissez-faire approach to memory permits more flexibility at the expense of fidelity to machine language programming— perhaps, ultimately, a smart tradeoff by Fred Matt, since Instructo programs can do so much more in so many fewer lines of code than the CARDIAC or LMC, thus holding the waning interest of teenagers itching to slip to the TRS-80 computer lab next door to play *Pyramid 2000* or *Android Nim*.

building a calculator program

Just like we did last section with the CARDIAC and LMC, we will introduce programming the device by building a functional calculator (keeping in mind, of course, that even though the Instructo will run the program we'll still have to do the arithmetic by hand).

While we have much more freedom with data types than we did with the CARDIAC or LMC, we also have a number of structural considerations to review.

1. Before running any program— or entering any inputs needed in the program itself— we need to press the "RESET/CLEAR" switch, which does the following: Sets the Program Step Indicator and Index Counter to zero; sets both registers to zero; sets both compare units to equals signs; and clears both outputs. The consequence of this is that every program must begin in storage location 00.

---

* Input A, though, is designed to take only integers, while Input B can process numbers or strings.

2. Only storage locations 90 to 99— recall, called the "Main Storage"— are read/write; the remaining storage locations ("Program Storage") house the read-only code for the program.
3. There are a maximum of seven input slots (or cards, or cells, or whatever you wish to call them). One such input slot is designated Input A, and only accepts a single decimal number. The other input, Input B, takes at most six entries of any data type (numbers, words, symbols, you name it). The operator's manual explains that Input B's operation is akin to a magnetic tape reader.
4. The Program Step Indicator keeps track of the next instruction to execute, while the Index Counter functions as an optional readymade loop counter. (Recall that with the CARDIAC or LMC we used memory cells on an ad hoc basis for loop counters.) The Index Counter can be referenced at any time by replacing an instruction's storage location (SS) number with the letters IN. The number showing on the index counter then becomes the SS referenced; for example, a conditional jump with an operand of IN would result in a conditional jump to the storage location showing on the Index Counter.
5. The three Jump Switches, which can be set to zero or one in a pinch by using the SJ*# mnemonic, usually function as flags to alert the Instructo to some condition.
6. The operator's manual separates the opcode's mnemonic abbreviation and operand by a comma. For instance, an unconditional jump to storage location 43 would appear as JUMP, 43. Nonetheless, the syntax in this guide will dispense with the extraneous comma, thus making JUMP 43 the equivalent valid instruction.

With all that out of the way, let's write a program that will add or multiply up to four numbers at a time.[*] We'll only be using Input B here: the first piece of data to enter is either the word ADD or MULTIPLY; then, enter in up to four numbers; and, finally, enter a terminating character (so the program knows when to stop retrieving input from Input B), which for us will be 0. Notice that the program code is discontinuous, skipping over quite a few lines (storage locations). We will be making use of some of those empty storage locations later.

Instructo Program No. 3-1: Adding and Multiplying

LINE	MNEMONIC	SS/IN	COMMENTS
00	ENIB	90	Enter information from input B into storage location 90 (this input, right now, must either be the word ADD or MULTIPLY).
01	LDRB	90	Load Register B with the contents of storage location 90.

---

[*] Since we can also subtract by entering as input negative numbers, a subtraction-specific routine will not be covered here.

02	CPRB	16	Compare the contents of Register B with storage location 16.
03	JBEQ	30	Jump to storage location 30 if Register B matches storage location 16 (i.e., go to the ADD subroutine).
04	CPRB	17	Compare the contents of Register B with storage location 17.
05	JBEQ	40	Jump to storage location 40 if Register B matches storage location 16 (i.e., go to the MULTIPLY subroutine).
*	*	*	
15	0		Read-only data storage.
16	ADD		Read-only data storage.
17	MULTIPLY		Read-only data storage.
*	*	*	
	ADD SUBROUTINE		
30	ENIB	91	Enter first number from Input B into storage location 91.
31	ENIB	92	Enter next number from Input B into storage location 92.
32	LDRB	92	Load Register B with the contents of storage location 92.
33	CPRB	15	Compare the contents of Register B with storage location 15 (which is zero).
34	JBEQ	38	Jump to storage location 38 only if the numbers in Register B and storage location 92 are equal.
35	ADDB	91	Add the number in Register B with the number in storage location 91.
36	STRB	91	Store the contents of Register B into storage location 91.
37	JUMP	31	Unconditional jump to storage location 31.
38	PROB	91	Output the number in storage location 91.
39	STOP	00	End the program.
	MULTIPLY SUBROUTINE		
40	ENIB	91	Enter first number from Input B into storage location 91.
41	ENIB	92	Enter next number from Input B into storage location 92.
42	LDRB	92	Load Register B with the contents of storage location 92.
43	CPRB	15	Compare the contents of Register B with storage location 15 (which is zero).
44	JBEQ	48	Jump to storage location 48 only if the numbers in Register B and storage location 92 are equal.
45	MULB	91	Multiply the number in Register B with the number in storage location 91.
46	STRB	91	Store the contents of Register B into storage location 91.
47	JUMP	41	Unconditional jump to storage location 41.
48	PROB	91	Output the number in storage location 91.
49	STOP	00	End the program.

The adding and multiplying subroutines are, in effect, identical (save for the arithmetic operation itself).

Expanding the calculator to handle division is easy using the Instructo's DIV* mnemonic, which writes the quotient and remainder to different registers. We will only set the program to perform a single division, however, meaning only three items need to be entered into Register B: the word DIVISION, the dividend, and the divisor, in that order. The quotient will appear in Register B, and the remainder in Register A.

The instruction set also allows for straightforward exponentiation courtesy of the EXP* mnemonic, so let us take advantage of that, too. Again, three items need to be entered into Register B: the word POWER, the base, and the exponent, in that order.

To allow for division and exponentiation, add the following lines of code to the calculator program listed above:

Instructo Program No. 3-1 (cont.): Division and Exponentiation

LINE	MNEMONIC	SS/IN	COMMENTS
06	CPRB	18	Compare the contents of Register B with storage location 18.
07	JBEQ	50	Jump to storage location 50 if Register B matches storage location 18 (i.e., go to the DIVIDE subroutine).
08	CPRB	19	Compare the contents of Register B with storage location 19.
09	JBEQ	61	Jump to storage location 61 if Register B matches storage location 19 (i.e., go to the POWERS subroutine).
*	*	*	
18	DIVIDE		Read-only data storage.
19	POWER		Read-only data storage.
20	THE QUOTIENT IS		Read-only data storage.
21	THE REMAINDER IS		Read-only data storage.
*	*	*	
	DIVIDE SUBROUTINE		
50	ENIB	91	Enter first number from Input B into storage location 91.
51	ENIB	92	Enter next number from Input B into storage location 92.
52	LDRB	91	Load Register B with the contents of storage location 91.
53	DIVB	92	Divide the number in Register B by the number in storage location 92, placing the quotient in Register B and the remainder in Register A.
54	STRB	93	Store the contents of Register B into storage location 93.
55	STRA	94	Store the contents of Register A into storage location 94.
56	PROB	20	Output the value in storage location 20.
57	PROB	93	Output the number in storage location 93.
58	PROB	21	Output the value in storage location 21.
59	PROB	94	Output the number in storage location 94.
60	STOP	00	End the program.
*	*	*	
	POWERS SUBROUTINE		
61	ENIB	91	Enter first number from Input B into storage location 91.
62	ENIB	92	Enter next number from Input B into storage location 92.
63	LDRB	91	Load Register B with the contents of storage location 91.
64	EXPB	92	Take the value in Register B to the power stored in storage location 92.

65	STRB	93	Store the contents of Register B into storage location 93.
66	PROB	93	Output the number in storage location 93.
67	STOP	00	End the program.

So far what we've done isn't particularly impressive; in fact, seeing how simple it is to perform division— obtaining a quotient and a remainder— makes our complex decimal-to-binary program we constructed last section rather trivial to write on the Instructo (and is therefore left as an exercise to the reader).[*]

The Instructo offers us more a powerful instruction set than either the CARDIAC or LMC, so let's see if we can take advantage of it to add a function to our calculator program that would be infeasible to code on the CARDIAC or LMC.

The sine of an angle, using right triangle trigonometry, is equal to the ratio of the angle's opposite side to the hypotenuse (i.e., the side opposite the right angle). But using right triangles in effect restricts our inputs; we want to be able to input any angle (in radians, though, not degrees), and immediately get a value for that angle's sine function.

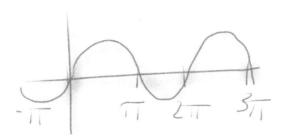

Instead, then, we'll need to use the first several terms of the Taylor series expansion of the sine function (which has an infinite number of terms), calculated by summing successively higher-order derivatives evaluated at a specific point. Here is the Taylor series for the sine function evaluated at $x = 0$:[†]

$$\sin x = x - \frac{x^3}{3!} + \frac{x^5}{5!} - \frac{x^7}{7!} + \cdots$$

---

[*] Better yet, write a program to have the Instructo relay binary conversions of most single-digit decimal numbers— but use the Jump Switches to print output.

[†] Technically, this expansion of terms is called a Maclaurin series, since we're evaluating the Taylor series at $x = 0$.

which, evaluating the factorials (the integers with exclamation marks in the denominators), can be rewritten as

$$\sin x = x - \frac{x^3}{6} + \frac{x^5}{120} - \frac{x^7}{5040} + \cdots$$

While it might be clever to have a subroutine evaluating the factorials in the series (the Instructo's operator's manual contains a program for producing factorials), there's really no need to here: since we will only be using the first four terms of the expansion to obtain our sine value, the numbers 6, 120, and 5040 are the only evaluated factorials we need— and they won't change. The powers in the numerators won't change, either. And four terms is enough to obtain an accurate value of sine to around four or five digits after the decimal point, as long as we stick with angle values in radians relatively close to 0.[*]

To get the sine of an angle using the Instructo's calculator program, two items will need to be entered into Input B: the word SIN (which is the mathematical abbreviation for sine) and the angle (in radians). To convert an angle in degrees to radians, use the following formula:

$$rad = deg \cdot \frac{\pi}{180}$$

So, for instance, 60 degrees is 1.0472 radians.

To allow for evaluating the sine function, we're going to have to free up some memory. Let's get rid of the exponentiation function. Add/replace the following lines of code to the calculator program listed above:

Instructo Program No. 3-1 (cont.): Sine Function

LINE	MNEMONIC	SS/IN	COMMENTS
08	CPRB	22	Compare the contents of Register B with storage location 22.
09	JBEQ	61	Jump to storage location 61 if Register B matches storage location 61 (i.e., go to the SINE subroutine).
*	*	*	
22	SIN		Read-only data storage.
23	3		Read-only data storage.
24	5		Read-only data storage.
25	7		Read-only data storage.
26	6		Read-only data storage.
27	120		Read-only data storage.

---

[*] Remember, our Maclaurin series evaluates the Taylor series at x = 0, so the farther we stray from an input of 0 radians, the more terms of the series we will need to evaluate to maintain a commensurate level of decimal-expansion precision.

| 28 | | 5040 | | Read-only data storage. |
| * | | * | * | |

SINE SUBROUTINE

61	ENIB	90	Enter radian angle measure from Input B into storage location 90.
62	LDRB	90	Load Register B with the contents of storage location 90 (i.e., the radian angle measure).
63	EXPB	23	Take the value in Register B to the power stored in storage location 23.
64	STRB	91	Store the contents of Register B into storage location 91.
65	LDRB	91	Load Register B with the contents of storage location 91.
66	DVDB	26	Divide the contents of Register B by the number in storage location 26, storing the (decimal) quotient in Register B.
67	STRB	92	Store the contents of Register B into storage location 92.
68	LDRB	90	Load Register B with the contents of storage location 90 (i.e., the radian angle measure).
69	EXPB	24	Take the value in Register B to the power stored in storage location 24.
70	STRB	91	Store the contents of Register B into storage location 91.
71	LDRB	91	Load Register B with the contents of storage location 91.
72	DVDB	27	Divide the contents of Register B by the number in storage location 27, storing the (decimal) quotient in Register B.
73	STRB	93	Store the contents of Register B into storage location 93.
74	LDRB	90	Load Register B with the contents of storage location 90 (i.e., the radian angle measure).
75	EXPB	25	Take the value in Register B to the power stored in storage location 25.
76	STRB	91	Store the contents of Register B into storage location 91.
77	LDRB	91	Load Register B with the contents of storage location 91.
78	DVDB	28	Divide the contents of Register B by the number in storage location 28, storing the (decimal) quotient in Register B.
79	STRB	94	Store the contents of Register B into storage location 92.
80	LDRB	90	Load Register B with the contents of storage location 90 (i.e., the radian angle measure).
81	SUBB	92	Subtract the number in Register B from the number in storage location 92.
82	ADDB	93	Add the number in Register B to the number in storage location 93.
83	SUBB	94	Subtract the number in Register B from the number in storage location 94.
84	STRB	95	Store the contents of Register B into storage location 95.
85	PROB	95	Output the number in storage location 95.
86	STOP	00	End the program.

If in Input B you type in SIN and 1.0472, which is approximately 60 degrees, the output will be 0.866022496, which is reasonably close to the exact value[*] of the sine of 60 degrees:

---

[*] Which can be found by using a 30-60-90 special triangle along with the right triangle definition of the sine function.

$$\frac{\sqrt{3}}{2} \approx 0.86602540378\ldots$$

So our sine algorithm works. And yet— something about it feels a little unseemly. Even though the powers of the terms in the numerator start at 3 and always go up by two for each successive term (starting with the second term, of course), while the factorials in the successive denominators also follow a predictable pattern, suppose we wanted to evaluate the Taylor series to five terms, or six, or seven— using the methods in the program above, we would have to keep adding more and more lines of code until we (quickly) run out of memory. Surely there's a better way: set up subroutines to handle the numerator (which will have exponentiation), the denominator (which will have repeated multiplication, courtesy of a factorial calculator), and the division itself; in addition, the signs of the quotients need to be alternated as they are summed together. It's a fairly difficult program to write, so I'll leave it as a reader's challenge— but not before I offer several hints in the form of programs.

First, listed below is a program to calculate a factorial. For instance, enter in 5, and 120 will result as output.

Instructo Program No. 3-2: Factorials

LINE	MNEMONIC	SS/IN	COMMENTS
00	ENIA	90	Store your input factorial into storage location 90.
01	LDRA	17	Load the number 1 into Register A.
02	LDRB	17	Load the number 1 into Register B.
03	ADDB	17	Add 1 to the number in Register B.
04	STRB	92	Store the number in Register B into storage location 92.
05	MULA	92	Multiply the number in Register A by the number in storage location 92.
06	STRA	91	Store the number in Register A into storage location 91.
07	LDRB	90	Load the number in storage location 90 into Register B.
08	SUBB	17	Subtract 1 from the number in Register B.
09	STRB	90	Store the number in Register B into storage location 90.
10	CPRB	17	Compare Register B with the number in storage location 17, which is 1.
11	JBEQ	15	Jump to storage location 15 if the Register B compare unit shows an equals sign.
12	LDRA	91	Load the number in storage location 91 into Register A.
13	LDRB	92	Load the number in storage location 92 into Register B.
14	JUMP	03	Unconditional jump to storage location 03.
15	PROA	91	Output the number in storage location 91.
16	STOP		End the program.
17	1		Read-only data storage.

Next, the program below takes two inputs— a base and the number of total terms— and takes the base to ever-increasing odd powers (starting from 1), alternating signs with each exponentiation, and then summing all of the resultant terms (however many you wished to see). The final result, along with each calculated term, is shown as output.

Instructo Program No. 3-3: Sum Alt-Sign Terms of Increasing Odd Powers

LINE	MNEMONIC	SS/IN	COMMENTS
00	ENIA	90	Store your input base into storage location 90.
01	ENIB	92	Store your input number of total terms into storage location 92.
02	LDRA	90	Load the number in storage location 90 into Register A.
03	LDRB	31	Load the number in storage location 31 (which is 1) into Register A.
04	STRB	91	Store the number in Register B into storage location 91.
05	EXPA	91	Take the value in Register A to the power in storage location 91.
06	STRA	98	Store the number in Register B into storage location 98.
07	LDRA	92	Load the number in storage location 92 into Register A.
08	DIVA	32	Divide the number in Register A by the number in storage location 32, which is 2; place the remainder in Register B.
09	CPRB	30	Compare Register B with the number in storage location 30, which is 0.
10	JBEQ	14	Jump to storage location 14 if the Register B compare unit shows an equals sign.
11	LDRA	98	Load the number in storage location 98 into Register A.
12	MULA	33	Multiply the number in Register A by the number in storage location 33, which is -1 (making the term negative).
13	STRA	98	Store the number in Register A into storage location 98.
14	LDRA	93	Load the number in storage location 93 into Register A.
15	ADDA	98	Add the number in storage location 98 to the number in Register A.
16	PROB	98	Output the number in storage location 98.
17	STRA	93	Store the number in Register A into storage location 93.
18	LDRA	91	Load the number in storage location 91 into Register A.
19	ADDA	32	Add the number in storage location 32 (which is 2) to the number in Register A.
20	STRA	91	Store the number in Register A into storage location 91.
21	LDRA	92	Load the number in storage location 92 into Register A.
22	SUBA	31	Subtract 1 from the number in Register A.
23	STRA	92	Store the number in Register A into storage location 92.
24	CPRA	30	Compare Register A with the number in storage location 30, which is 0.
25	JAEQ	28	Jump to storage location 28 if the Register A compare unit shows an equals sign.
26	LDRA	90	Load the number in storage location 90 into Register A.
27	JUMP	05	Unconditional jump to storage location 05.
28	PROA	93	Output the number in storage location 93.
29	STOP		End the program.

30	0	Read-only data storage.
31	1	Read-only data storage.
32	2	Read-only data storage.
33	-1	Read-only data storage.

To write that Taylor series program, all that's left to do is fuse together the two programs above, adding a routine to divide the results of the second program by the first. You might want to hold off on working on that program, though, until you read the next subsection.

assembly language programming

If we could detach ourselves from worrying about the storage locations with each instruction, programming the Instructo would be simpler. Assembly language allows us to do just that— but we'll have to lay several ground rules first.

Unlike the CARDIAC manual, which briefly sketches out the syntax of a CARDIAC assembly language, the Instructo's operator's manual does no such thing— assembly language isn't even mentioned.* Therefore, the criteria we set forth below aren't necessarily "canonical," in the sense of what Fred Matt would have done. Nonetheless, we press on.

1. We will introduce the DAT mnemonic for initializing variables: both the read-only kind and the read-write kind. All DAT instructions should be at the end of the program.
2. We can define as many read-only variables as memory allows— for example, initializing the read-only variable "ten" to a numeric value of 10, or the read-only variable "status" to a string value of "YOU WIN."
3. But we are permitted to initialize only ten read-write variables (which, behind the scenes, are arbitrarily assigned storage locations 90 to 99). Initialize more than ten read-write variables, and things get messy as new variables start replacing older ones.
4. Except with the DAT instruction, absolutely no numeric operands— like ADDA 10, which seems like it should add 10 to the value in Register A— are permitted. Instead, initialize a read-only variable called "ten" and then use "ten" as the operand: ADDA ten.
5. The operand IN always functions as a call to the value showing on the Index Counter.

---

* Recall that the CARDIAC manual scaffolds a scrap of assembly code to help the reader understand high-level languages like FORTRAN. Although in theory possible to code, in practice writing a high-level language with the Instructo is infeasible because of the limited available memory, analogous to the CARDIAC and the LMC.

6. The mnemonics STOP and NOOP have no associated operands; they are used solely as standalone instructions.

First, we will write an Instructo program is machine language; then, we'll translate the same program into assembly.

Consider a "hailstone number," which is also referred to as a $3n+1$ sequence. A hailstone number can be defined as follows by using function notation:

$$f(n) = \begin{cases} \dfrac{n}{2} \text{ if } n \text{ is even} \\ 3n+1 \text{ if } n \text{ is odd} \end{cases}$$

For instance, consider the number 6. Plugged into the function, 6 is even, so we must divide it by 2, obtaining 3. But 3 is an odd number, so we take 3×3+1 = 10. Then we see that 10 is even, so we divide by 2: our result is 5. If we continue, we find that 6 takes 8 iterations to land at 1.[*] Notice that we used a recursive procedure to find the answer: 8.

A hailstone number program must calculate the number of iterations a starting integer "seed" takes to reach 1 by following the recursive function rule just described. Here is such a program, along with a flowchart for it, taking as input a single value into Input A.

Instructo Program No. 3-4: Hailstone Numbers

LINE	MNEMONIC	SS/IN	COMMENTS
00	ENIA	93	Store your (integer) input hailstone number into storage location 93.
01	LDRA	93	Load the hailstone number into Register A.
02	DIVA	08	Divide the hailstone number by the number in storage location 08, which is 2; place the remainder in Register B.
03	CPRB	09	Compare Register B with the number in storage location 09, which is 0.
04	INDA	10	Add to the Index Counter the number in storage location 10, which is 1.
05	JBEQ	11	Jump to storage location 11 if the Register B compare unit shows an equals sign.
06	JBGT	17	Jump to storage location 17 if the Register B compare unit shows a greater than sign.
07	3		Read-only data storage.
08	2		Read-only data storage.
09	0		Read-only data storage.
10	1		Read-only data storage.

---

[*] The Collatz conjecture states that any natural number will, with a finite number of iterations, eventually settle down to 1.

"IF VALUE IS EVEN" SUBROUTINE

11	STRA	93	Store the contents of Register A into storage location 93.
12	PROA	93	Output the number in storage location 93.
13	LDRA	93	Load the value in storage location 93 into Register A.
14	CPRA	10	Compare Register A with the number in storage location 10, which is 1.
15	JAEQ	23	Jump to storage location 23 if the Register A compare unit shows an equals sign.
16	JUMP	01	Unconditional jump to storage location 01.

"IF VALUE IS ODD" SUBROUTINE

17	LDRA	93	Load the value in storage location 93 into Register A.
18	MULA	07	Multiply the value in Register A by the value in storage location 07.
19	ADDA	10	Add the value in Register A by the value in storage location 10, which is 1.
20	STRA	93	Store the contents of Register A into storage location 93.
21	PROA	93	Output the number in storage location 93.
22	JUMP	01	Unconditional jump to storage location 01.

"IF VALUE IS ONE" SUBROUTINE

23	PROB	26	Output the number in storage location 26.
24	PROB	27	Output the number in storage location 27.
25	STOP	00	End the program.
26	SEE INDEX COUNTER		Read-only data storage.
27	FOR HAILSTONE #		Read-only data storage.

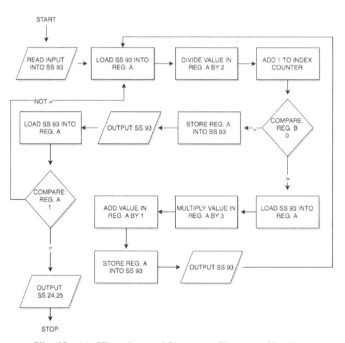

Fig. No. 11. Flowchart of Instructo Program No. 3-4.

Entering in 6 into Input A results in the following output:

Output A	Output B
3	SEE INDEX COUNTER
10	FOR HAILSTONE #
5	
16	
8	
4	
2	
1	

The Index Counter at the end of the run will show 08, meaning there were eight iterations until a 1 was reached, including the 1 (as is shown in Output A above).

A corresponding Instructo assembly language hailstone program might look like the following:

Instructo Program No. 3-5: Hailstone Numbers Redux

LABEL	MNEMONIC	OPERAND	COMMENTS
start	ENIA	hailstone	Stores in the (integer) user input as the variable "hailstone."
begin	LDRA	hailstone	Loads the value of "hailstone" into Register A.
	DIVA	two	Divides the value in Register A by 2, placing the remainder into Register B.
	CPRB	zero	Compares the value in Register A with 0.
	INDA	one	Adds 1 to the Index Counter.
	JBEQ	even	If the value in Register A is equal to zero, then jump to the "even" label.
	JBGT	odd	If the value in Register A is greater than zero, then jump to the "odd" label.
even	STRA	hailstone	Store the value in Register A into "hailstone."
	PROA	hailstone	Output the value of "hailstone."
	LDRA	hailstone	Loads the value of "hailstone" into Register A.
	CPRA	one	Compares the value in Register A with 1.
	JAEQ	end	If the value in Register A is equal to one, then jump to the "end" label.
	JUMP	begin	Unconditional jump to the "begin" label.
odd	LDRA	hailstone	Loads the value of "hailstone" into Register A.
	MULA	three	Multiplies the value in Register A by 3.
	ADDA	one	Adds 1 to the value in Register A.
	STRA	hailstone	Store the value in Register A into "hailstone."
	PROA	hailstone	Output the value of "hailstone."
	JUMP	begin	Unconditional jump to the "begin" label.
end	PROB	text1	Output the value of "text1."
	PROB	text2	Output the value of "text2."
	STOP		End the program.
three	DAT	3	Initialize a read-only variable.
two	DAT	2	Initialize a read-only variable.
one	DAT	1	Initialize a read-only variable.
zero	DAT	0	Initialize a read-only variable.
text1	DAT	SEE INDEX COUNTER	Initialize a read-only variable.
text2	DAT	FOR HAILSTONE #	Initialize a read-only variable.
hailstone	DAT		Initialize a read-write variable.

The assembly program is still long and moderately complicated to work through, but simpler than the associated machine language program; labels help to organize the subroutines, and variable names bring a clarity that references to storage locations lack. And as I'm sure you're well aware by now, when writing any computer code, clarity is an ally we can never afford to forsake.

a (pseudo)random number generator

Computers do not produce genuine random numbers; rather, *pseudorandom* numbers, or numbers that appear to be random but are in fact deterministic, are generated instead.

Producing pseudorandom numbers is usually accomplished with an iterative, perhaps even recursive, mathematical function. To begin the process, an initial random number seed must be input— either by the user directly, or by the computer indirectly (possibly obtained from its internal clock or some other physically unpredictable process)— that initializes the string of random numbers which will, given enough time, repeat in sequence. John von Neumann was one of the first mathematicians to arrive at a pseudorandom number generator algorithm.

One of most widely used pseudorandom number generator algorithms is called the "linear congruential generator." Given some initial seed value X,

$$X_{N+1} = (aX_N + c) \bmod m$$

where a and c are the multiplier and the increment, respectively, and m is the modulus. Note that the modulo operator $\bmod$ divides the quantity $(aX_N + c)$ by m but ignores the quotient in favor of the remainder. Also note that the "next" term $N+1$ of the generating sequence is obtained by feeding in the result of the "current" term N into the formula (as denoted by the subscripts of the X's).

Let's create an Instructo assembly program that loosely utilizes the linear congruential generator algorithm to produce a string of "random" numbers. Our program will take five integer inputs: into Input A, an initial seed; and into Input B, a value for the multiplier a, the increment c, the modulus m, and the number of numbers to be generated, in that order.

Instructo Program No. 3-6: Pseudorandom Number Generator

LABEL	MNEMONIC	OPERAND	COMMENTS
start	ENIA	seed	Stores in the (integer) user input as the initial seed.
	ENIB	a	Stores in the (integer) user input as the multiplier, "a."
	ENIB	c	Stores in the (integer) user input as the increment, "c."
	ENIB	m	Stores in the (integer) user input as the modulus, "m."
	ENIB	number	Stores in the (integer) user input as the number of numbers to generate.
	INDL	number	Sets the Index Counter to the value of the "number" variable.
begin	LDRA	seed	Loads the value of the "seed" into Register A.
	MULA	a	Multiplies the value in Register A by the value of the "a" variable.
	ADDA	c	Adds the value of the "c" variable to the number in Register A.

	DIVA	m	Divides the value in Register A by the "m" variable, placing the remainder in Register B.
	STRB	seed	Stores the value in Register B into the "seed" variable.
	PROB	seed	Output the value of "seed."
	INDS	one	Decrement the Index Counter by one unit.
	JIZE	end	If Index Counter equals zero, then jump to the "end" label.
	JUMP	begin	Unconditional jump to the "begin" label.
end	STOP		End the program.
one	DAT	1	Initialize a read-only variable.
m	DAT		Initialize a read-write variable.
seed	DAT		Initialize a read-write variable.
number	DAT		Initialize a read-write variable.
a	DAT		Initialize a read-write variable.
c	DAT		Initialize a read-write variable.

After running the program several times, it will quickly be apparent that the convincingness of the "randomness" of the numbers generated depends entirely on the set of user inputs. Some sets of inputs produce strings of numbers that betray an obvious pattern; other inputs simply result in outputs of the same number repeated over and over again. It takes trial and error to find the best mix of inputs.

For a further challenge, reproduce this linear congruential generator on the CARDIAC or LMC.

one final program: a game of single-pile nim

In the earliest days of electronic computing, having your machine play games like Nim was all the rage, mostly because programming sophisticated games like chess was either infeasible or impossible. Unsurprisingly, courtesy of a internet search you can easily find multiple versions of the game Nim for the LMC and the CARDIAC; an extended section on coding the game is in the CARDIAC manual, which focuses on single-pile Nim (i.e., each player can only remove objects from a single, common pile)[*] with a key exception: in each turn, the player and her opponent cannot remove the same number of objects; so, for instance, if you remove three objects— in the case of the CARDIAC manual, pebbles are used— your opponent cannot also remove three.[†] We will dispense with this exception when writing our very simple single-pile Nim program— but for the Instructo, which, as far as I can tell, has no such program extant.

---

[*] Nim doesn't have to be restricted to one pile.

[†] An edifying, but exceptionally difficult, programming exercise: translate the CARDIAC manual's machine language Nim programs (there are two) into assembly code.

Let us set the rules of our Nim game as follows: there is a single pile of marbles in front of you and your opponent, which will be the computer. When your turn comes, you are permitted to take one, two, or three marbles from the pile; the computer is also allowed to take one, two, or three marbles. The loser of the game is the player who's left with one or zero marbles remaining in the pile.

The pile will be stocked with 17 marbles. You will go first, writing your first move into Input A before running the program; the computer will output its move in Output A, and the number of remaining marbles will be shown in Output B. Then you'll have to enter your next move as the second input of Input B,[*] and the game will continue until a player is faced with zero or one marble left in the pile.

Instructo Program No. 3-7: Single-Pile Nim

LABEL	MNEMONIC	OPERAND	COMMENTS
start	LDRA	seventeen	Load the number 17 into Register B.
	STRA	marbles	Store the number 17 into the "marbles" variable.
loop	ENIA	yourmove	Store in the user input as the variable "yourmove."
	SUBA	yourmove	Subtract the value in Register A by the value in variable "yourmove."
	STRA	marbles	Store the value in Register A into the variable "marbles."
	CPRA	one	Compare the value in Register A to 1.
	JAEQ	endwin	If the value in Register A is equal to one, then jump to the "endwin" label.
	JALT	endwin	If the value in Register A is less than one, then jump to the "endwin" label.
	LDRB	four	Load the number 4 into Register B.
	SUBB	yourmove	Subtract the variable "yourmove" from 4.
	STRB	compmove	Store the value in Register B into the variable "compmove."
	PROA	compmove	Output the value of "compmove" (to Output A).
	SUBA	compmove	Subtract the value in Register A by the value in variable "compmove."
	STRA	marbles	Store the value in Register A into the variable "marbles."
	PROB	marbles	Output the value of "marbles" (to Output B).
	CPRA	one	Compare the value in Register A to 1.
	JAEQ	endlose	If the value in Register A is equal to one, then jump to the "endlose" label.
	JALT	endlose	If the value in Register A is less than one, then jump to the "endlose" label.

---

[*] If you're using the Instructo emulator presented later on to play the game, you'll have to turn the debugging mode on and stop the emulator (by pressing the Escape key) each time the computer completes a move so that you can enter *your* next move into Input A, and then restart the program. By the way, the same start-stop process is also necessary if running a machine language CARDIAC/LMC Nim game using the emulator provided in this guide (it's also necessary in any other game requiring dynamic inputs by the user, for that matter).

	NOOP		No operation— if running this program on the emulator found later on in this guide, here's the place to stop (hit Escape several times) and enter in your next move into Input A. Then restart the program.
	JUMP	loop	Unconditional jump to the "loop" label.
endwin	PROA	wintext	Output the value of "wintext" (to Output A).
	PROB	wintext	Output the value of "wintext" (to Output B).
	STOP		End the program
Endlose	PROA	losetext	Output the value of "losetext" (to Output A).
	PROB	Losetext	Output the value of "losetext" (to Output B).
	STOP		End the program.
Wintext	DAT	YOU WIN!	Initialize a read-only variable.
Losetext	DAT	YOU LOSE!	Initialize a read-only variable.
One	DAT	1	Initialize a read-only variable.
Four	DAT	4	Initialize a read-only variable.
Seventeen	DAT	17	Initialize a read-only variable.
Marbles	DAT		Initialize a read-write variable.
Yourmove	DAT		Initialize a read-write variable.
Compmove	DAT		Initialize a read-write variable.

Run the program, and after a couple of rounds you'll discover that the computer is unbeatable; the greatest Nim strategy players in the world couldn't win. (Because the computer can't lose, some of the code, such as the text "YOU WIN" and the subroutine to check if you beat the computer, is superfluous.) The reason is simple: Since the computer's move is the difference between your previous move and the number four, the sum of your move and the computer's move for each turn is also four— and four, when divided into seventeen (the initial count of marbles in the pile), leaves a remainder of one.

So here is your final Instructo reader's challenge: Write a game of single-pile Nim that isn't unwinnable, but not too easy, either; i.e., have the computer put up a good fight. Hint: having the computer anticipate a user's possible moves— and searching for multiples of a certain number n after marbles have been removed by leveraging the values of remainders— might be a fruitful avenue for exploration. Also consider having the computer occasionally remove a (pseudo)random number of marbles, just to keep things unpredictable.

SECTION 4.

instructional models and the future

The CARDIAC, LMC, and Instructo do not exist in a vacuum; rather, each influenced and/or was influenced by other instructional models of electronic computers.

the tutac

The TUTorial Automatic Computer (TUTAC) was the brainchild of Theodore G. Scott, detailed in his lengthy book from 1962 entitled *Basic Computer Programming*, along with a follow-up two years later called *Computer Programming Techniques*. Scott's books were only two of the many Doubleday[*] TutorText series books (they were volume number 7 and 21, respectively), which were designed to teach a specific specialty by using a "gamebook," a work of fiction that permitted the reader some narrative interactivity through choice-making (perhaps the best known gamebooks are the later *Choose Your Own Adventure* series). Other gamebooks in the TutorText series included not only math and computer science titles like *Practical Mathematics*, *The Arithmetic of Computers*, *Adventures in Algebra*, *The Slide Rule*, *Trigonometry: A Practical Course*, but also works on Shakespeare (*Understanding Shakespeare: Macbeth*), poetry (*The Meaning of Modern Poetry*), the Constitution and the law (*The American Constitution*, *Practical Law: A Course in Everyday Contracts*, and *Advanced Bidding*), and games (*The Elements of Bridge* and *The Game of Chess*), among other subjects.

Both the CARDIAC and the TUTAC are decimal-based machines. But compared to the CARDIAC, the TUTAC is a monster: 10,000 memory cells, thirteen opcodes (two digits apiece), an accumulator, an unsigned ten-digit register, and fifteen-digit-long machine language instructions that include: a special digit denoting the sign (0 for positive, 1 for negative), the four-digit address (since there are 10,000 memory cells: numbers 0000 to 9999), and even space for a ten-digit-long integer.

---

[*] Later reissued by English Universities Press and Macdonald.

Let us briefly examine the thirteen opcodes. There's opcode 00 (Stop), which halts the execution of the program, as well as four arithmetic opcodes: 60 (Add), 61 (Subtract), 62 (Multiply), and 63 (Divide). There are also two opcodes to "bit" shift by a single digit: 08 (Shift Right) and 09 (Shift Left). The unconditional jump is opcode 30 (Unconditional Transfer), along with two opcodes for conditional jumps: 31 (Transfer If Zero) and 32 (Transfer If Positive). Three opcodes remain, and all deal with input/output: 15 (Store), 25 (Write), and 50 (Copy). The TUTAC's input/output paradigm, unsurprisingly for its year of creation, is punched-card input and typewriter output.

Any program run on the TUTAC begins at memory cell 0000. Interestingly, there is an "overflow light" that turns on to stop the program if the accumulator tries to process a number too large.

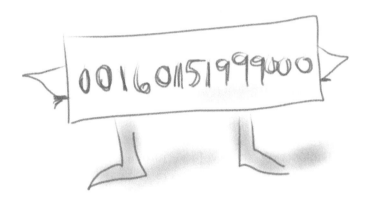

In retrospect, it's all a bit too much; the TUTAC makes the fully loaded Instructo look like a stripped-down model. The CARDIAC and LMC took the best of the TUTAC— whether intentionally or not— and instead offered a simple package to the computer neophyte. Programming the CARDIAC and the LMC requires sets of three-digit words; programming the TUTAC necessitates fifteen-digit words, taking into account opcodes, addresses, signs, and types of decimal data.

But the TUTAC gamebook— which was called a "multiple-choice, branching" book at the time— was widely promoted as an early computer training tool, as this critical review[*] from the periodical *Computers and Automation* (March 1963) attests:

> Obviously, the author [Theodore G. Scott] can only be responsible for proving conclusively that learners do learn TUTAC programming. The training director as a purchaser must be sure the TUTAC programming is

---

[*] https://archive.org/stream/bitsavers_computersA_7421473/196303_djvu.txt

readily transferable to programming for the specific computer being used. The use of pseudo or fictional computers is a practice in education environments where machines are often not available, rather than in business and industry where specific hardware and software techniques exist.... The author deserves an A for Effort.

An A for effort, certainly. But certainly not for ease of use.

the tis-100

There is a direct line between the TUTAC of 1962 and the TIS-100 of 2015. Whereas the TUTAC wrapped the study of computing in a decidedly low-tech gamebook, the TIS-100, or Tessellated Intelligence System, is a one-of-a-kind videogame in the puzzle genre released by Zachtronics.[*] In it, the player is presented with a corrupted early 1980s-style computer, complete with a monochromatic text-based display; the player's assignment is to sort through the processor errors and perform challenging programming tasks by writing assembly code.

The game is quite complex; there are multiple accumulators, registers, and processing nodes to consider— and the paths that data (in the form of inputs and outputs) take around the console's nodes are traced by arrows. Writing effective assembly code is your only strategy to tackling the corrupted regions of the TIS-100; for example, mnemonics like MOV come into play: MOV 10,ACC [format: source, destination] moves the value of 10 into the accumulator. There are arithmetic mnemonics available, such as ADD and SUB, along with NEG, which flips the sign of the value in the accumulator. And the JMP mnemonic performs an unconditional jump to a labeled block of code (a label must be have a colon appended; for instance, LOOP: and START: are valid labels), while the JEZ mnemonic conditionally jumps depending on if the value zero is present in the accumulator. A step-through debugger helps you to better navigate the complexities on the console. You are scored on cycle, node, and instruction counts, and your statistics are compared online with other players'. Twenty-first century programming fun, courtesy of virtual computer similar to a TRS-80.

the paperclip computer

The 1967 book *How to Build a Working Digital Computer* by Edward Alcosser, James P. Phillips, and Allen M. Wolk[†] brought blueprints of the digital age to intrepid tinkerers and students all across America. Using nothing more than items you might find around your house— light bulbs, thread,

---

[*] http://www.zachtronics.com/tis-100/

[†] https://archive.org/details/howtobuildaworkingdigitalcomputer_jun67

wire, and paperclips (hence the "paperclip computer" moniker)— you can have your own electronic, binary, and digital computer up and running. As Windell Oskay explains, "The instructions include a read-only drum memory for storing the computer program (much like a player piano roll), made from a juice can, with read heads made from bent paper clips. A separate manually-operated 'core' memory (made of paper-clip switches) is used for storing data."*

To program the paperclip computer, an octal (base 8) code, rather than a binary code, was utilized; octal numbers decreased the space requirements for the programming instructions. The opcodes included ADD, SUB, STO (store from the accumulator to the core memory), SHR (shift right), SHL (shift left), TRA (transfer), JUP (unconditional jump), COJ (conditional jump), RIN (read in an input into the accumulator), and RUT (display accumulator value in output).

But, although the book is a fantastic introduction to a computer's operation and logic, the build instructions are extremely complex and are probably too much for most to tackle, evidenced by the fact that in the fifty years since publication only several fully operational models are known to have been built from the ground up by hobbyists. One of those, called the Emmerack, was made in 1972 by two teenagers, Mark Rosenstein and Kenny Antonelli. As an adult Rosenstein recalls,

> The drum memory was a paint can, that a patient of my father, who worked in a factory that produced paint cans, had grabbed before they put on the handles....
>
> A program was written in a machine language, which we called Wierd (Written Idiomatic Expressions Reduced Drastically), and punched onto a file folder which was wrapped around the paint can, held on by electrical

---

* http://www.evilmadscientist.com/2013/paperclip/

tape. A row of paper clips made contact with the drum memory and caused instructions to be lighted up. (This mechanism was not that reliable, and something more springy than paper clips was needed).[*]

So much for the paperclips.

Technically, however, the paperclip computer nickname was given to a mass-produced computer thought to have been clandestinely *built off of* the blueprints presented in *How to Build a Working Digital Computer*: the Comspace Computer Trainer-650 (CT-650), built by Irving Becker in the late sixties. For $1,000 (around $7,000 in today's dollars), you could buy a fully assembled CT-650, complete with an input unit (using switches to take input: accepting decimal, converting to binary), an output unit (converting binary to decimal, and displaying outputs using light bulbs), an arithmetic unit, a control unit, and memory storage— but no paperclips. Programs were available to order through the Arkay Program Library.

the little men

The most "hacked" paper computer is the LMC. Stuart Madnick's instructional paradigm has been modified, and emulated electronically, numerous times. The article "A Crowd of Little Man Computers: Visual Computer Simulator Teaching Tools" by William Yurcik and Hugh Osborne[†] (from February 2001) details some of the LMC variants.

There's the *Postroom Computer*, which extends the LMC with the following instruction set architectures: a zero-address stack-based machine; a one-address accumulator-based machine (akin to the CARDIAC and unmodified LMC); a two-address machine, with a source and a destination operand; and a three-address machine, with two source operands and a single destination operand. The zero-address stack-based machine includes two opcodes for manipulating stacks: PSH (push to the stack) and POP (pop from the stack). The two- and three-address models add the MOV (move value) and MEA (move effective address) opcodes. Both direct and indirect addressing are possible, along with the use of a set of registers.

Other models include the *Son-of-LMC*, which is an extension of the LMC instruction set that permits subroutine calls and returns (as does the more recent Little Man Computer on the Raspberry Pi; the Raspberry Pi is a small, easily modifiable circuit board designed for students and hobbyists interested in computers), and the LMC-1 and LMC-2, which present fully fledged graphical visualizations of the fetch/execute cycle of the Little Man.

The paper "The 'Little Man Storage' Model," by Larry Brumbaugh and William Yurcik (from June 2005), details a significant modification of the

---

[*] http://www.apparent-wind.com/mbr/emmerack.html

[†] http://citeseerx.ist.psu.edu/viewdoc/download?doi=10.1.1.1.4370&rep=rep1&type=pdf

LMC paradigm permitting the simulation of disk/tape storage. The authors posit a storage paradigm for the LMC consisting of the following storage device components: three cylinders (0, 1, and 2), four tracks (and two platters) per cylinder (0, 1, 2, and 3), and eight areas per track (0, 1, 2, 3, 4, 5, 6, and 7), with each area storing 512 bytes of data. The addressing scheme for the areas is set as xyz, where x is the cylinder, y the platter, and z the area; a disk cache is unavailable. Input/output operations can be executed on disk areas using the LMS (Little Man Storage paradigm), hewing much closer to the way memory management is performed on modern computers.

the paper computer as stage play

As a retired professor in the department of computer science at the University of São Paulo, Brazil, Valdemar W. Setzer has an extensive background in computers, education, and their intersection. Setzer earned a doctorate in applied mathematics from the Polytechnic School of Universidade de São Paulo in 1967, and is perhaps best known for a late-nineties article entitled "The Obsolescence of Education," which argued for a humane education, free of the "retrograde characteristics of education," divorced of the reflexive use of the computer, especially with preadolescents: "With those and many other considerations, I reached the conclusion that a young person should never use a computer before puberty; due the necessary self-control, the ideal age being around 17." And this is despite the temptation of educators to use the technology at hand (all the more so two decades after the article's publication):

> I think that one of the main reasons for people advocating the use of computers in education (besides surrendering to the intense propaganda of hardware and software producers) is the fact that they are extremely attractive. In a certain way, this fact should be used as a warning: a machine attracting more than a human being is an aberration. This is due, as I exposed, mainly to teachers not considering the changes which occurred in the human being during the 20th century, and continue to teach in obsolete ways in relation to those changes. Thus, classes are boring, excessively abstract, learning is enforced, etc. What is happening is that, instead of correcting these mistakes at their origin, that is, changing the teachers' mentality, the computer is introduced as a kind of crutch. It is believed that this way education is modernized. Basically, this is a deceit, because the computer attracts through its "cosmetic" or electronic game effects and not through the contents or personal relationships.[*]

When recently interviewed about the article, Setzer doubled down on his conclusions— warning especially of the dangers of internet and its promotion of anti-social behavior:

---

[*] http://www.ime.usp.br/~vwsetzer/obsol-eng.html

> There are dozens of other negative effects of the Internet, but these two I consider definitive and irrefutable: anything that is addictive and dangerous should not be used by children and adolescents. Therefore, parents, schools and teachers who give incentives to the use of the Internet are definitively damaging their children and students. This puts even more emphasis on my main argument: teaching should be more humane, and not more technological.[*]

He also warned against introducing abstractions in mathematics until the high school grades.

How does any of this relate to paper computers? Well, when he was still a classroom teacher, Setzer had an intriguing idea.

> In 1976 I had the idea of introducing a visual activity to teach the basic notions of the internal structure of a computer through its machine language. I devised the "Paper Computer," which is a sort of theater play where the actors are students, simulating every device of the HIPO [from the Portuguese *computador HIPOtético*, or hypothetical computer] machine. Contrary to the latter, the instructions were not coded but written in extensive text for easy and immediate understanding. The activity, which takes 2 hours, uses about 20 students, who perform at the front of the class, so all other students may follow what is happening and give suggestions.[†]

The Paper Computer itself was assembled from twenty students, each wearing a piece of carton, with a string attached, designating his or her computational role (CPU, accumulator, instruction pointer, and so forth). Instructions were designated by means of storage positions; the instruction set included inputting in a number (03), a conditional jump (06), and halting the program (12)— a very standard but basic instruction set. Through teacher and student directives, the "computer" was capable of running a program. It's like a Little Man Computer, but where the entire architecture is made of *actual* little men (and little women).

The Paper Computer play served as a low-tech (or tech-free) kinesthetic educational activity teaching how computers functioned, and helped to "put machines in their proper context [and]... provide an *understanding of the world*," according to Setzer.

---

[*] http://www.truth-out.org/speakout/item/26131-valdemar-w-setzer-on-the-obsolescence-of-education

[†] https://www.ime.usp.br/~vwsetzer/paper-comp.html

the (German) wdr paper computer

Several years after the Instructo Paper Computer was released, another paper computer arrived on the scene in West Germany. Called the WDR Paper Computer, "WDR" stands for Westdeutscher Rundfunk Köln, a federal public-broadcasting institution in Germany.

In 1983, a television program called ComputerClub,[*] hosted by German actor and television personality Wolfgang Back, hit the airwaves in West Germany. Though the program proved popular, few viewers in the country owned or had access to computers at the time. So Back and German scientist and entrepreneur Ulrich L. Rohde created their paper computer which, according to Back, sold at least 300,000 units after being presented on the television show.

The teaching aid, designed to simulate the functioning of a computer and the writing of low-level machine language programs, was similar to the CARDIAC in look and operation; it too resembled a board game to the uninitiated, although the CARDIAC's presentation was much cleaner. The WDR Paper Computer had an arithmetic unit, memory cells, registers, and a program counter; a ballpoint pen, instead of the bug of the CARDIAC, functioned as the program pointer. Only five commands, having the following mnemonics, were available: JMP (jump directly to a specified address), ISZ (check to see if a specified register is equal to zero), INC (increment a specified register), DEC (decrement a specified register), and STP (stop the computer). As a program ran, the "bits" of data (not actually bits, since the computer operated in decimal, not binary), which were represented by wooden matches— the WDR Paper Computer's central conceit— were alternatively placed in and taken out of the eight available registers. Compare the WDR's five commands to the ten of the CARDIAC's instruction set: the German paper computer didn't even have the ADD and SUB of the CARDIAC yet could still perform basic arithmetic operations.[†] The WDR Paper Computer was about as simplified as you could get.

Early this century, Wolfgang Back wrote an emulator of his paper computer in Visual Basic.[‡] Indeed, Back was a rare bird: actor, television host, programmer, computer science educator, Renaissance man. The Know-How Computer, a slight variant of the WDR Paper Computer, is still used today to teach the basics of computer programming to students in Namibia.

---

[*] https://www.youtube.com/watch?v=G5-H2T1chZ4

[†] https://marian-aldenhoevel.de/wp-content/themes/generatepress-child/papercomputer/occ_know_how.jpg

[‡] http://www.wolfgang-back.com/knowhow_home.php

and the rest

There have been several other notable efforts with computers made of paper that should at least be briefly touched upon.

A Paper Enigma Machine. Alan Turing used the electronic, digital, programmable Colossus computer at Bletchley Park to crack Germany's Enigma machine during World War II. Using a complex of ciphers, it is possible to construct an Enigma machine made of common household objects, à la the paperclip computer, including paper.[*]

The Paper Processor. Created by Saito Yutaka in 2010, the two-bit binary Paper Processor,[†] a two-dimensional schematic map of a computer laid out on a poster board, has multiple registers but only three instructions: increment register by one, jump if there's not an overflow, and halt. Interestingly, the Paper Processor models Harvard architecture, rather than von Neumann, making it the only known paper computer to do so.

The One Instruction Set Computer (OISC). Designed by Peter Crampton, a former student of Drexel University professor Brian L. Stuart, the OISC[‡] is similar to the CARDIAC: it decimal-based and has one hundred memory cells, with a somewhat analogous instruction set. But each instruction is six digits long, split into two digits apiece: source operand, destination operand, and target address. Data stored into memory cell 98 is printed as output; a jump to memory cell 99 results in the machine halting.

Electronic Paper Computers. The idea emerged in the anything-is-possible late nineties: disposable paper computers, costing only several dollars at most. The brainchild of Jim Willard, these paper computers took their inspiration from greeting cards— containing cheap microchips embedded in disposable cardboard— and were enough of a novel idea that *Wired* magazine featured an article about them on April 1, 2000 (title: "Computers, Real Cheap!"),[§] which, in retrospect, seems like an elaborate April Fools' joke. At the time, Willard's goal was to sell his paper computers to the IRS, turning their paper forms into dynamic, interactive pulp. Needless to say, things didn't quite unfold that way.

---

[*] http://www.howtogeek.com/115767/make-your-own-paper-craft-enigma-machine-diy-project/

[†] https://sites.google.com/site/kotukotuzimiti/

[‡] https://www.cs.drexel.edu/~bls96/oisc/

[§] http://www.wired.com/2000/04/paper/

the future: a quantum paper computer

In 1965, Gordon Moore predicted that the number of transistors compressed onto a microchip will double approximately every 24 months— meaning that computer processing power will double roughly every two years. Although Moore's law has aligned closely to reality for well over half a century, it cannot continue to remain in force indefinitely; at some point, transistors will not be able to shrink any further (such as to an atomic scale, if that's even possible).

Which brings us to quantum computers, a still-theoretical construct that ditches transistors in favor of performing operations on quantum bits (termed "qubits") of data using such exotic physics concepts as "superposition" and "entanglement." We're used to the bits of classical computing, which can assume a value of either 0 or 1. Quantum computing uses qubits, which can also be 0 or 1, except they can also be 0 and 1 *simultaneously*— this is the notion of superposition (and how Schrödinger's cat somehow manages to be alive and dead at the same time). The more qubits you have access to, the more your processing power grows; in fact, it grows exponentially.

If quantum computing is even possible— and that's a big if, since the physical conditions necessary to maintain a qubit in superposition may not be feasible— then calculations that binary digital electronic computers require inordinate lengths of time to complete might take quantum computers mere nanoseconds. Moore's law would have been shown to be too conservative of a prediction. Online security systems, such as those that keep your banking and credit card information safe from hackers (thanks to complex encrypting algorithms involving prime factorization), would all, in principle, be rendered nonfunctional thanks to a quantum factoring algorithm developed by MIT professor Peter Shor.

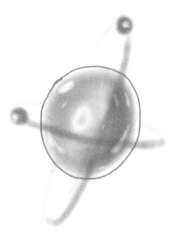

Perhaps in the future, quantum computing will indeed replace classical computing. Back in the sixties, when classical computing was emerging and Gordon Moore was first proposing his take on computing power, the CARDIAC offered high school students a tactile means to learn about computers' functionality, sans expensive hardware. Might a quantum paper computer be in the offing— a sort of "CARDIAQUC," or CARDboard Illustrative Aid to QUantum Computation— to teach quantum computing to that first generation of young folks only vaguely familiar with the still-too-expensive-to-procure quantum hardware? The answer lies in the future, too.

PART B.

pulp and digital emulation

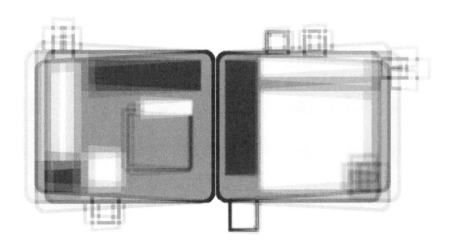

SECTION

the pink paper computer

Running paper computer programs presents you with a bit of a dilemma: namely, what device should you use? Should you try to track down a rare CARDIAC, or an even rarer Instructo? Or, instead, should you attempt to recreate these paper-and-cardboard devices by printing online images and literally cutting and pasting? Or maybe you should make use of emulator software?[*] Perhaps simply "running" them in your head will suffice?

If, however, it feels to you "inauthentic" to run a CARDIAC, LMC, or Instructo program on anything other than strips of paper and cardboard, but you are either unable to get ahold of an original pulp device or unwilling to attempt to reproduce (using online schematics) one, an alternative is offered here: fashioning a brand new paper computer.

Called the Pink Paper Computer, or PPC for short, the device, once assembled, can run CARDIAC, LMC, and Instructo programs with little fuss but a fair amount of pencil-and-paper calculation along with a dose of dexterity— but not much more than the erstwhile pulp devices required.

Although an actual, *color* Pink Paper Computer would have made this (physical) book prohibitively expensive to produce and purchase, consider the "Pink" moniker a political shot across the bow of those who would continue to ignore the interests of females in computer science as well as deny their enduring contributions.

assembling the ppc

The PPC has a number of components, and thus a handful of steps required for full assembly. These steps are presented below; the components are printed on the pages at the end of this section.

1. Using scissors, cut out all the components.

---

[*] Multiple versions of which are in subsequent sections of this guide.

2. Back the two big components, called the *mainboard* and the *storage cells* (each of which takes up a full book page), with cardboard.
3. After backing, use an utility knife to cut out the outlines of the "windows" of the mainboard (each such window contains a ✂ icon in its center).
4. The remaining components to cut out are wheels, strips, and struts. The two sets of wheels with numbers around their circumferences must be stacked on top of one another (but not pasted together), like this:

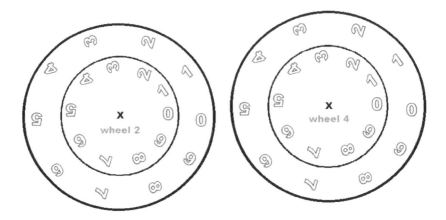

Next, wheels should be attached at their centers (x's mark the spots) with brass round-head fasteners behind the mainboard; the fasteners need to slice right through the mainboard and permit the wheels to rotate. The light semicircular outlines on the mainboard will help you to position the wheels:

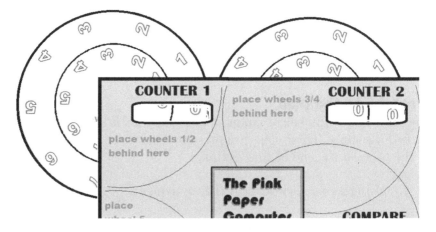

5. Place wheels 5 and 6 in their correct positions, using brass round-head fasteners inserted through the mainboard to permit rotation.

6. Fasten, using tape, the struts onto the back of the mainboard. Two pieces of tape should be used for each strut; the tape should be placed at the ends of the struts, perpendicularly, to allow the strips to slide easily through them. Two struts should be placed behind the input column on the mainboard, and two struts should be placed behind the output column. In addition, one strut should be taped behind the accumulator. See the diagram below for the strut placements, bearing in mind that the struts are positioned *behind* the mainboard, not in the front.

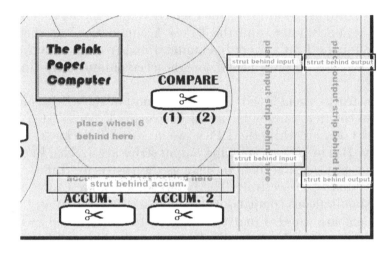

7. Finally, insert the input, output, and accumulator strips into the mainboard.

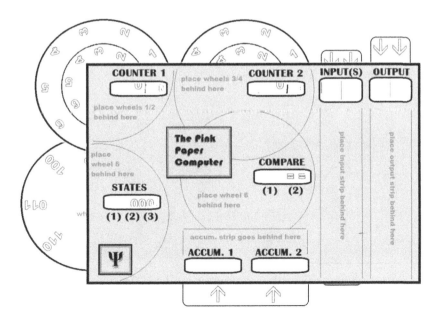

And also have the separate sheet of storage cells handy, along with a pencil and eraser.

Your PPC is now ready to run machine language programs, but there are key differences in how to use the pulp device depending on the kind of program run.

running machine language programs on the ppc

Running machine language programs on either the CARDIAC or the LMC is relatively straightforward. For the CARDIAC, use the PPC's Counter 1 as the bug/instruction counter[*] and the PPC's Accumulator 1 as the accumulator. Likewise for the LMC: use only Counter 1 and Accumulator 1. Make sure to write lightly, using only pencil, on the accumulator strip and the storage cells.

The Instructo, though, will make use of most of the PPC's components. The PPC's two counters can double as the Instructo's Program Step Indicator and the Index Counter, while the PPC's two accumulators can stand in as Registers A and B. The PPC's input and output strips are divided in two by vertical lines; this split permits them to serve as the Instructo's Inputs A and B and Outputs A and B, respectively. The Instructo's Compare Unit— of which there are two simultaneous comparisons possible— is functional on the PPC by using the Compare wheel. Finally, the Instructo's three Jump Switches, each of which displays 0 or 1— is handled by using the binary States wheel on the PPC. The PPC's storage cells have spaces for 100 storage locations, which is just enough for the Instructo.

Expanding the PPC to run TUTAC programs, on the other hand, is most definitely best left as an exercise to the intrepid reader.

---

[*] Although they are technically not synonymous, here, for convenience, we consider them functionally the same: as both denoting the "current" cell/instruction.

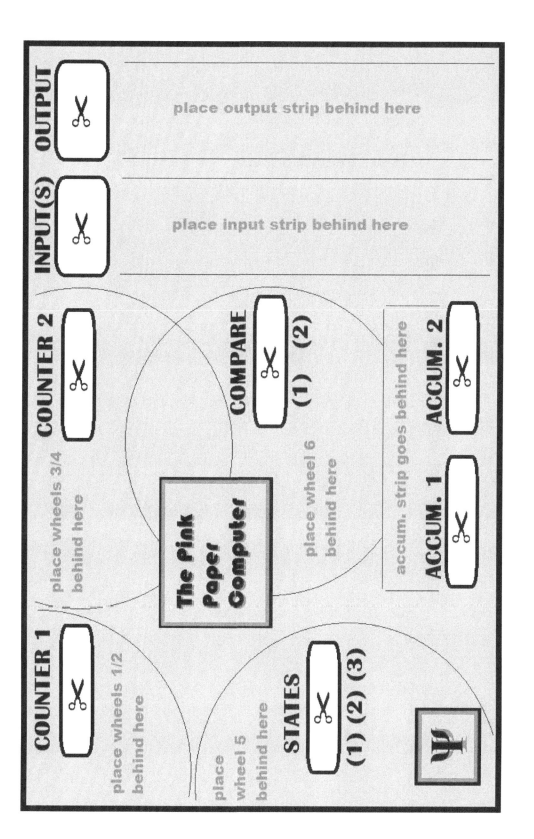

The Pink Paper Computer

STORAGE CELLS

00	17	34	51	68	85
01	18	35	52	69	86
02	19	36	53	70	87
03	20	37	54	71	88
04	21	38	55	72	89
05	22	39	56	73	90
06	23	40	57	74	91
07	24	41	58	75	92
08	25	42	59	76	93
09	26	43	60	77	94
10	27	44	61	78	95
11	28	45	62	79	96
12	29	46	63	80	97
13	30	47	64	81	98
14	31	48	65	82	99
15	32	49	66	83	
16	33	50	67	84	

make sure to write lightly, with pencil only, inside the storage cells

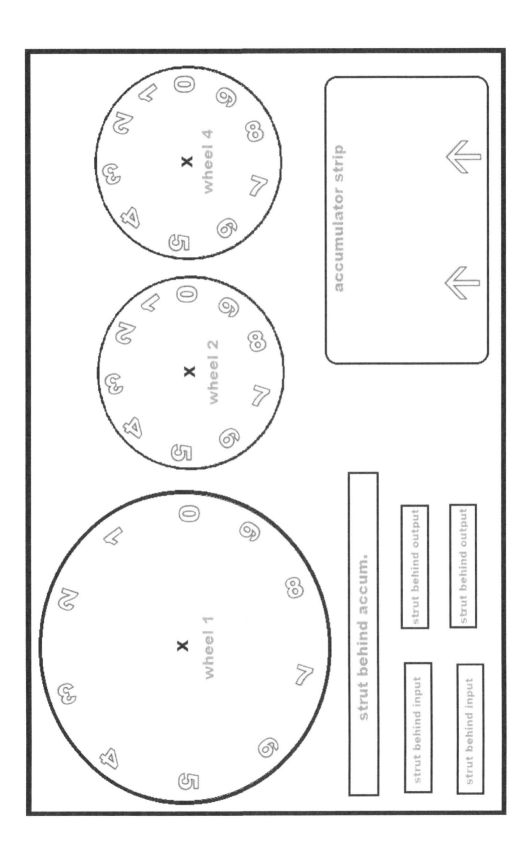

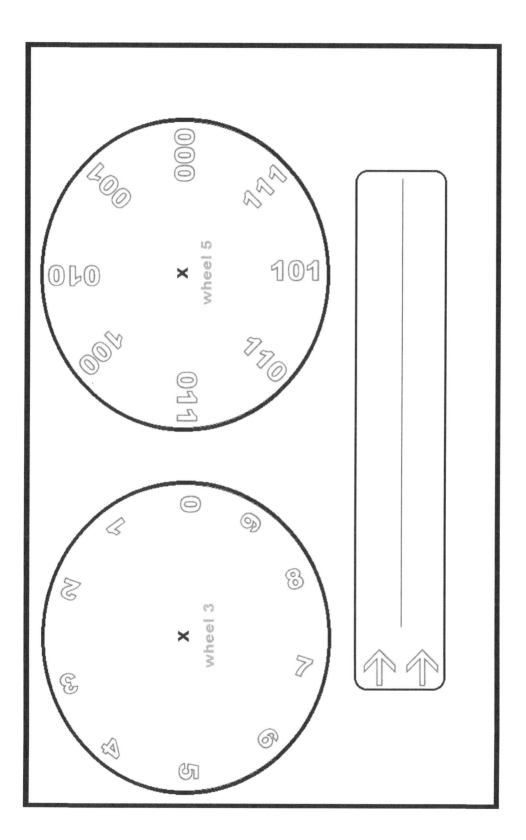

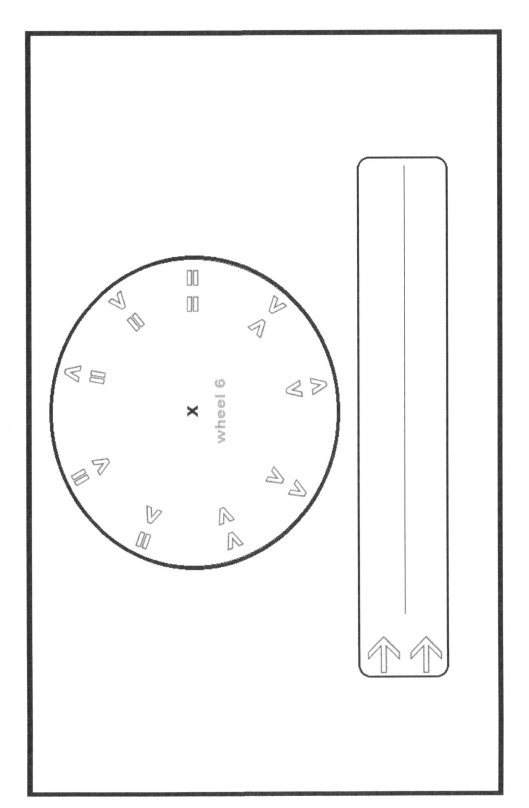

SECTION

a cardiac-little man emulator

Although it is true that programming an electronic, virtual version of a paper computer seems to render the whole enterprise beside the point, remember that paper computers were designed to be educational devices. In that sense, virtual electronic paper computers, henceforth referred to as *emulators*, satisfy that same educative function. The "paper" part of these paper computers is mostly incidental to their functionality as teaching tools; in the sixties and seventies, these computers could have been made partly of plastic or metal or even stone with no loss of educative function. In fact, the paper computer didn't really need to be physically constructed at all, as long as the computer operator knew her machine language inside and out. Programs could be executed entirely in the mind, with perhaps a bit of assistance courtesy of paper and pencil. The CARDIAC manual almost says as much, noting at the tail end of its relaying an especially knotty counting program that "Tilt won't be necessary to run this program through CARDIAC. We can spot a serious drawback simply by looking it over." Although, admittedly, the more complex the program, the more difficult such a run in the mind's eye will be.

So, since we're knee-deep in the *Age of the Computer* (as the CARDIAC manual terms it), let's take most of the physical tedium out of running a CARDIAC/LMC machine language program by creating a Microsoft Excel macro[*] to run it for us. Such macros, which are built using Visual Basic code embedded in spreadsheets, can automate otherwise tedious, repetitive tasks.

---

[*] An Excel spreadsheet, with its dynamic rows and columns, is a natural fit for a paper computer emulator (or two, or three). Several years ago, when beginning research on the CARDIAC, I encountered an interesting post on CPU design by Al Williams on the Dr. Dobb's: The World of Software Development website (www.drdobbs.com). In the post, after lamenting that purchasing a classroom set of actual CARDIACs with which to teach computer programming was logistically infeasible, Williams provides descriptions of and links to CARDIAC and Little Man Computer emulators, both of which were made with Excel. He also mentions how he himself first learned about computers: by using the TUTAC.

Since the CARDIAC and the LMC are similar in design and approach, we'll construct a *single* macro to emulate both paper computers. Although there are a number of emulators available online for the CARDIAC and the LMC— written in a wide variety of computer languages and operating systems— none, so far as I can find, package the two together.

getting started

There are a number of versions of Excel available, so the macro-creation instructions will be sufficiently general.

First, you'll need to make sure that you have the "Developer" tab in the Excel "Ribbon" (the strip of options right below the menu) available. Go to "File," then "Options," then "Customize Ribbon." Finally, check off the "Developer" option. Even if you don't have a "Ribbon," you'll need to make sure that the "Developer" option is selected in order to proceed.

Open up a new Excel spreadsheet, and then save it with the filename "CARDIAC-LMC." Make sure your workbook is "Macro-Enabled" when saving.

setting up the worksheet

First, right click the worksheet's tab name, type "CARDIAC-LMC," and then click back inside the worksheet.

We want the emulator to display which paper computer is in use. Bold cell A1, and then type in the following:

```
=IF(L28=0,"'CARDIAC' developed by David Hagelbarger","'LITTLE MAN COMPUTER' developed by Stuart Madnick")
```

The CARDIAC and LMC use different terminology for their von Neumann architecture. Memory cells in the CARDIAC are called mailboxes in the LMC; the accumulator in the CARDIAC is the calculator in the LMC (although some versions of the LMC in fact do call the accumulator an accumulator); and so on.

Let's, then, account for these differences of language between the two paper computers. Bold cells A3, A23, E23, P2, R2, and T2. In cell A3, type:

```
=IF(L28=0,"MEMORY CELLS","MAILBOXES")
```

In cell A23, type:

```
PROGRAM/INSTRUCTION COUNTER
```

In cell E23, type:

```
=IF(L28=0,"ACCUMULATOR","CALCULATOR")
```

In cell P2, type:

=IF(L28=0,"INPUT","INBOX")

In cell R2, type:

=IF(L28=0,"OUTPUT","OUTBOX")

In cell T2, type:

=A3

That completes all worksheet formulas' setup. All remaining worksheet setup is cosmetic (and non-dynamic); the inputs into remaining cells will be text or numbers, rather than formulas (which always begin with an equals sign).

We will start by setting up a space for the instruction register (or program counter) and accumulator (or calculator). In cell A25, type 0. The format of the 0, however, needs leading zeros, so it looks like this: 000. In order to make that change, right click the cell and select "Format Cells." Once there, select the "Custom" option and, in the text box labeled "Type:" type in 000 and press the "OK" button. Next, draw a border around the instruction register by highlighting cells A23 to C26, right clicking the shaded region, selecting the "Format Cells" option again, clicking the "Border" tab, and then clicking the "Outline" button.

Now we will set up the accumulator. In cell E25, type a plus sign (+). Cell F25 needs to display 0000. Type 0 into cell F25, then alter the format of the cell, nearly the same as you did with cell A25 but with four zeros this time. Finally, the accumulator needs a border; highlight cells E23 to G26 to create it.

The correct layout for the instructor register (program counter) and accumulator (calculator) is shown below.

Next up: a "control panel," showing options for paper computer emulation. Start by drawing borders around cells L29, L29, L30, L31, and L32. Then, enter 0's in each of these cells except L30— in which the number 1 should be entered. Next, right-justify cells K28 to K32. And, finally, enter the following text— whose purpose will be explained in the next subsection— into the right-justified cells:

the paper computer unfolded • 173

	Use which computer? (CARDIAC = 0, LMC = 1)	0
	Pauses? (0 = No, 1 = Yes)	0
	See Steps? (0 = No, 1 = Yes)	1
	Convert mnemonics (numeric codes) to machine language before running? (0 = No, 1=Yes)	0
	Convert machine language (numeric codes) to mnemonics before running? (0 = No, 1=Yes)	0

On to the memory cells: first, holding down the Control (Ctrl) key,[*] highlight the cells A5 to A21, C5 to C21, E5 to E21, G5 to G21, I5 to I21, and K5 to K19. Center-justify and bold the highlighted cells, and change the cell display format to show 00. Click off the highlighted cells. Type 001 into cell B5, and 800 into cell L19— these are the two default machine language instructions for the CARDIAC. Then draw a border around the cells A3 to L21. Next, type in all of the address numbers for the memory cells, as shown in the following screenshot:

	A	B	C	D	E	F	G	H	I	J	K	L
1	'CARDIAC' developed by David Hagelbarger											
2												
3	MEMORY CELLS											
4												
5	00	001	17		34		51		68		85	
6	01		18		35		52		69		86	
7	02		19		36		53		70		87	
8	03		20		37		54		71		88	
9	04		21		38		55		72		89	
10	05		22		39		56		73		90	
11	06		23		40		57		74		91	
12	07		24		41		58		75		92	
13	08		25		42		59		76		93	
14	09		26		43		60		77		94	
15	10		27		44		61		78		95	
16	11		28		45		62		79		96	
17	12		29		46		63		80		97	
18	13		30		47		64		81		98	
19	14		31		48		65		82		99	800
20	15		32		49		66		83			
21	16		33		50		67		84			

There's still a bit more to do. Using the same format as the memory cells, type the numbers 1 to 24, one number apiece, into cells O4 to O27. And put a border around cells O2 to P27. Make sure you screen matches the screenshot on the top of the next page.

---

[*] If you're using an Apple, hold down the ⌘ (Command/Apple) key instead.

O	P
	INPUT
	01
	02
	03
	04
	05
	06
	07
	08
	09
	10
	11
	12
	13
	14
	15
	16
	17
	18
	19
	20
	21
	22
	23
	24

Our emulator is going to be able to convert from mnemonics to machine language and back again, both for the CARDIAC and the LMC. Let's fashion a workspace for entering mnemonics. In cell U2, type "MNEMONIC." In cell V2, type "ADDRESS." And in cell W2, type "COMMENTS." Make sure all three words are bolded. Then— using the same format as the memory cells— from cells T3 to T102, type the numbers 00 to 99, one number per cell. Make sure that cells V3 to V102 have the same format as the memory cells as well.

To match the CARDIAC's two default machine language instructions using mnemonics, type "INP" into cell U3 and 01 into cell V3, along with "JMP" into cell U102 and 00 into cell V102. Wrapping a border around cells T2 to W102, as shown below, effectively completes the setup of the worksheet.

R	S	T	U	V	W
		MEMORY CELLS	MNEMONIC	ADDRESS	COMMENTS
OUTPUT		00	INP	01	
		01			
		02			
		03			
		04			
		05			
		06			
		07			
		08			
		09			

coding the macro

Although what follows is not meant to be a primer on VBA (Visual Basic for Applications), or on coding in general, we will spill some ink discussing the reasoning behind parts of the code.

We'll first need to insert a button that will run our macro. Click on the "Developer" tab, select the "Insert" dropdown, and click the gray button. Draw a rectangular button, starting around cell I23; once you let go of the mouse button, a dialog box will pop up asking you which macro you'd like to assign to the button. Click "Cancel" for now, and instead right click the button and select "Edit Text." Type "START" and click somewhere outside of the button. Your worksheet should resemble this screenshot:

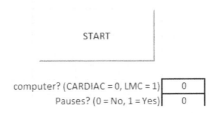

It's now time to insert the macro. Right click the "START" button, select "Assign Macro," and click the "New" button. Underneath the subroutine called `Sub Button1_Click()`, type the following:

```
'Run Program macro

Dim Rowct As Integer
Dim Columnct As Integer
Dim OutputRow As Integer
Dim InputRow As Integer
Dim OPCode As String
Dim Address As String
Dim Accumulator As Long
Dim LineofCode As String
Dim Code As String
Dim PreviousAddress As String
Dim FirstDigit As String
Dim SecondDigit As String
Dim RowTemp As Integer
Dim ColTemp As Integer
Dim AddressTemp As Integer
Dim CellValue As Integer
Dim AddrValue As Integer
Dim Computer As Integer
```

The `Dim` statements above declare important variables for later use.* Some variables are of type `String`, meaning that characters of any sort can be

---

* Note, however, that most variable declaration— and, thus, the `Dim` statement— is optional in Visual Basic.

stored in them, while others are of type Integer, which is self-explanatory. The Long variable allows for a more precise number than the Integer type.

Sprinkled throughout the code are comments— shown as phrases that begin with apostrophes— that have been added for clarity but which are ignored when the macro's run. For example, the next bit of code clears the worksheet of any previously entered mnemonics:

```
'Clear any previous mnemonics:
Range("U3:W102").Select
    Selection.ClearContents
```

Throughout the rest of the macro, we'll have to keep alternating between CARDIAC code and LMC code— the variable Computer will keep track: 0 = CARDIAC, 1 = LMC— as shown below:

```
Computer = Range("L28").Value        'Tells macro which paper computer we're
using--CARDIAC (value = 0) or LMC (value = 1)
```

Next, we will set default machine language instructions for the CARDIAC:

```
'If we're using the CARDIAC, then make sure to reset the machine in addresses
00 and 99:

If Computer = 0 Then

    Range("L19").Value = 800
    Range("B5").Value = 1
    Range("U3").Value = "INP"
    Range("V3").Value = 1
    Range("U102").Value = "JMP"
    Range("V102").Value = 0

End If
```

Here is a routine to convert mnemonic codes to machine language, if the option is selected. (Turning the option on means typing a 1 into cell L31.) CARDIAC code is first, LMC, second.

```
'Convert mnemonic codes to machine language (OPCodes/Addresses, also called
"numeric codes") in memory cells (if user requests it):

Counter1 = 0
RowTemp = 5
ColTemp = 66

If Computer = 0 And Range("L31").Value = 1 Then        'CARDIAC

    For A = 3 To 102

    AddressTemp = Range("V" & A)

    If Range("U" & A).Value = "INP" Then Range(Chr$(ColTemp) & RowTemp).Value
    = 0 + AddressTemp
```

```
    If Range("U" & A).Value = "CLA" Then Range(Chr$(ColTemp) & RowTemp).Value
    = 100 + AddressTemp
    If Range("U" & A).Value = "ADD" Then Range(Chr$(ColTemp) & RowTemp).Value
    = 200 + AddressTemp
    If Range("U" & A).Value = "TAC" Then Range(Chr$(ColTemp) & RowTemp).Value
    = 300 + AddressTemp
    If Range("U" & A).Value = "SFT" Then Range(Chr$(ColTemp) & RowTemp).Value
    = 400 + AddressTemp
    If Range("U" & A).Value = "OUT" Then Range(Chr$(ColTemp) & RowTemp).Value
    = 500 + AddressTemp
    If Range("U" & A).Value = "STO" Then Range(Chr$(ColTemp) & RowTemp).Value
    = 600 + AddressTemp
    If Range("U" & A).Value = "SUB" Then Range(Chr$(ColTemp) & RowTemp).Value
    = 700 + AddressTemp
    If Range("U" & A).Value = "JMP" Then Range(Chr$(ColTemp) & RowTemp).Value
    = 800 + AddressTemp
    If Range("U" & A).Value = "HRS" Then Range(Chr$(ColTemp) & RowTemp).Value
    = 900 + AddressTemp

    RowTemp = RowTemp + 1

    If RowTemp > 21 Then
        RowTemp = 5
        ColTemp = ColTemp + 2
    End If

    Next A

End If

If Computer = 1 And Range("L31").Value = 1 Then     'LMC

    For A = 3 To 102

    AddressTemp = Range("V" & A)

    If Range("U" & A).Value = "HLT" Or Range("U" & A).Value = "COB" Then
    Range(Chr$(ColTemp) & RowTemp).Value = 0
    If Range("U" & A).Value = "ADD" Then Range(Chr$(ColTemp) & RowTemp).Value
    = 100 + AddressTemp
    If Range("U" & A).Value = "SUB" Then Range(Chr$(ColTemp) & RowTemp).Value
    = 200 + AddressTemp
    If Range("U" & A).Value = "STA" Then Range(Chr$(ColTemp) & RowTemp).Value
    = 300 + AddressTemp
    If Range("U" & A).Value = "LDA" Then Range(Chr$(ColTemp) & RowTemp).Value
    = 500 + AddressTemp
    If Range("U" & A).Value = "BRA" Then Range(Chr$(ColTemp) & RowTemp).Value
    = 600 + AddressTemp
    If Range("U" & A).Value = "BRZ" Then Range(Chr$(ColTemp) & RowTemp).Value
    = 700 + AddressTemp
    If Range("U" & A).Value = "BRP" Then Range(Chr$(ColTemp) & RowTemp).Value
    = 800 + AddressTemp
    If Range("U" & A).Value = "INP" Then Range(Chr$(ColTemp) & RowTemp).Value
    = 901
    If Range("U" & A).Value = "OUT" Then Range(Chr$(ColTemp) & RowTemp).Value
    = 902

    RowTemp = RowTemp + 1

    If RowTemp > 21 Then
        RowTemp = 5
        ColTemp = ColTemp + 2
    End If
```

```
    Next A
End If
```

Here is a routine to convert machine language to mnemonic codes, if the option is selected. (Turning the option on means typing a 1 into cell L32.) Again, CARDIAC code is first, LMC, second.

```
'Convert OPCodes/Addresses (machine language, also called "numeric codes") to
mnemonic codes in memory cells (if user requests it):

Counter1 = 0
RowTemp = 5
ColTemp = 66

If Computer = 0 And Range("L32").Value = 1 Then      'CARDIAC

    For A = 3 To 102

    CellValue = Val(Range(Chr$(ColTemp) & RowTemp).Value)
    AddrValue = CellValue - Int(CellValue / 100) * 100

    If Int(CellValue / 100) = 0 Then
        Range("U" & A).Value = "INP"
        Range("V" & A).Value = AddrValue
    End If

    If Int(CellValue / 100) = 1 Then
        Range("U" & A).Value = "CLA"
        Range("V" & A).Value = AddrValue
    End If

    If Int(CellValue / 100) = 2 Then
        Range("U" & A).Value = "ADD"
        Range("V" & A).Value = AddrValue
    End If

    If Int(CellValue / 100) = 3 Then
        Range("U" & A).Value = "TAC"
        Range("V" & A).Value = AddrValue
    End If

    If Int(CellValue / 100) = 4 Then
        Range("U" & A).Value = "SFT"
        Range("V" & A).Value = AddrValue
    End If

    If Int(CellValue / 100) = 5 Then
        Range("U" & A).Value = "OUT"
        Range("V" & A).Value = AddrValue
    End If

    If Int(CellValue / 100) = 6 Then
        Range("U" & A).Value = "STO"
        Range("V" & A).Value = AddrValue
    End If

    If Int(CellValue / 100) = 7 Then
        Range("U" & A).Value = "SUB"
        Range("V" & A).Value = AddrValue
    End If
```

```
    If Int(CellValue / 100) = 8 Then
        Range("U" & A).Value = "JMP"
        Range("V" & A).Value = AddrValue
    End If

    If Int(CellValue / 100) = 9 Then
        Range("U" & A).Value = "HRS"
        Range("V" & A).Value = AddrValue
    End If

    If CellValue = 0 Then
        Range("U" & A).Value = ""
        Range("V" & A).Value = ""
    End If

    RowTemp = RowTemp + 1

    If RowTemp > 21 Then
        RowTemp = 5
        ColTemp = ColTemp + 2
    End If

    Next A

End If


If Computer = 1 And Range("L32").Value = 1 Then      'LMC

    For A = 3 To 102

    CellValue = Val(Range(Chr$(ColTemp) & RowTemp).Value)
    AddrValue = CellValue - Int(CellValue / 100) * 100

    If CellValue = 0 Then
        Range("U" & A).Value = "COB"
        Range("V" & A).Value = ""
    End If

    If Int(CellValue / 100) = 1 Then
        Range("U" & A).Value = "ADD"
        Range("V" & A).Value = AddrValue
    End If

    If Int(CellValue / 100) = 2 Then
        Range("U" & A).Value = "SUB"
        Range("V" & A).Value = AddrValue
    End If

    If Int(CellValue / 100) = 3 Then
        Range("U" & A).Value = "STO"
        Range("V" & A).Value = AddrValue
    End If

    If Int(CellValue / 100) = 5 Then
        Range("U" & A).Value = "LDA"
        Range("V" & A).Value = AddrValue
    End If

    If Int(CellValue / 100) = 6 Then
        Range("U" & A).Value = "BRA"
        Range("V" & A).Value = AddrValue
    End If
```

```
    If Int(CellValue / 100) = 7 Then
        Range("U" & A).Value = "BRZ"
        Range("V" & A).Value = AddrValue
    End If

    If Int(CellValue / 100) = 8 Then
        Range("U" & A).Value = "BRP"
        Range("V" & A).Value = AddrValue
    End If

    If CellValue = 901 Then
        Range("U" & A).Value = "INP"
    End If

    If CellValue = 902 Then
        Range("U" & A).Value = "OUT"
    End If

    If Range(Chr$(ColTemp) & RowTemp).Value = "" Then
        Range("U" & A).Value = ""
        Range("V" & A).Value = ""
    End If

    RowTemp = RowTemp + 1

    If RowTemp > 21 Then
        RowTemp = 5
        ColTemp = ColTemp + 2
    End If

    Next A

End If
```

Next, a number of variables are assigned initial values, and previous printed output is erased.

```
'Initialize the values of some important variables:

Accumulator = 0
OutputRow = 4
InputRow = 4
PreviousAddress = 0
FirstDigit = 0
SecondDigit = 0


'Clear any previous output:

Range("R4:R4000").Select
    Selection.ClearContents
```

Then, a number of variables are assigned initial values, and previous printed output is erased.

```
Set ProgramStep = Range("A25")    'Makes a variable called "ProgramStep" to
be the line number of the program

If ProgramStep < 0 Then ProgramStep = 0
If ProgramStep > 99 Then ProgramStep = 99
```

Because of the horizontal-vertical layout of the memory cells, quite a few machinations are required to ensure that the correct memory cell is being read. ASCII (American Standard Code for Information Interchange) codes are used for columns. But we can conveniently compress those machinations into a couple of function calls.

```
'Set correct initial row/column placement of bug:

Rowct = CorrectRow(Val(ProgramStep))
Columnct = CorrectColumn(Val(ProgramStep))
```

The accumulator is next set to zero:

```
'Set accumulator to zero

Range("F25").Value = Accumulator
Range("E25").Value = "+"
```

A Do...Loop, along with some basic error checking, is established to run machine language programs.

```
'The loop to run a CADIAC/LMC program starts here:

Do

'Reset correct initial row/column placement of bug:

Rowct = CorrectRow(Val(ProgramStep))
Columnct = CorrectColumn(Val(ProgramStep))

LineofCode = Format((Range(Chr$(Columnct) & Rowct).Value), "000")    'Gather
the three digits of the current line of code; the format function makes sure
there are three digits assigned

OPCode = Val(Mid(LineofCode, 1, 1))      'Get opcode

Address = Val(Mid(LineofCode, 2, 2))     'Get address

'If the opcode is 0, then there's an issue--the macro will think the opcode
doesn't exist; deal with this issue:

If Val(LineofCode) < 100 Then OPCode = 0 And Address =
Format(Val(LineofCode), "00")

'If Address registers as just a single digit, then assign it a leading zero:

If Val(Address) < 10 Then Address = Format(Val(Address), "00")
```

While the program is running, the memory cell corresponding to the program instruction is highlighted in yellow:

```
'Light up current memory cell (in yellow)

 Range(Chr$(Columnct - 1) & Rowct).Select
    With Selection.Interior
        .Pattern = xlSolid
        .PatternColorIndex = xlAutomatic
```

```
        .Color = 65535
        .TintAndShade = 0
        .PatternTintAndShade = 0
    End With
```

The emulator has a dedicated debugger. If the debugger is turned on (the option is selected by typing a 1 into cell L30), then before each line of code is run, an anticipatory dialog box appears.

```
'Debugger, Part 1: Next line is the pre-run line of code:

If Range("L30").Value = 1 Then MsgBox ("About to run: LineofCode = " & LineofCode & " and opcode is " & OPCode & " and Address is " & Address)
```

CARDIAC only: the following routine reads a single input into a designated memory cell (INP):

```
If Computer = 0 And OPCode = 0 And LineofCode <> "" Then

Code = "Read input card into cell ___ (mnemonic = INP)"

Set TempInput = Range("P" & InputRow)    'Gather the "current" input from the input card

Range(Chr$(CorrectColumn(Val(Address))) & CorrectRow(Val(Address))).Value = TempInput 'Copy input into the address listed

InputRow = InputRow + 1     'Increment the input card

End If
```

LMC only: the following routine reads a single input into the calculator (INP):

```
If Computer = 1 And OPCode = 9 And Address = 1 And LineofCode <> "" Then

Code = "Read inbox value into calculator ___ (mnemonic = INP)"

Set TempInput = Range("P" & InputRow)    'Gather the "current" input from the input card

Accumulator = TempInput    'Set accumulator to contents of the input card

If Accumulator > 9999 Then Accumulator = 0    'Accumulator is restricted to numbers with four digits

If Accumulator < -9999 Then Accumulator = 0    'Accumulator is restricted to numbers with four digits

Range("F25").Value = Accumulator    'Change value in accumulator

If Accumulator < 0 Then Range("E25").Value = "-" Else Range("E25").Value = "+"      'Checks to see whether accumulator value is negative or positive (or zero); show + or - on-screen

InputRow = InputRow + 1     'Increment the input card

End If
```

LMC only: the following routine reads the calculator's value and copies it to the outbox (OUT):

```
If Computer = 1 And OPCode = 9 And Address = 2 And LineofCode <> "" Then

Code = "Read calculator's value and copy to outbox ___ (mnemonic = OUT)"

Set TempInput = Range("F25")   'Gather the contents of the accumulator

Range("R" & OutputRow).Value = TempInput    'Print accumulator contents to output card

OutputRow = OutputRow + 1   'Increment the output card

End If
```

CARDIAC and LMC: the following routine makes use of the value present in the accumulator (or calculator) (CLA/LDA):

```
If (Computer = 0 And OPCode = 1) Or (Computer = 1 And OPCode = 5) And LineofCode <> "" Then

If Computer = 0 Then Code = "Clear accumulator and add into it the contents of cell ___ (mnemonic = CLA)"

If Computer = 1 Then Code = "Replace the value in the calculator with the contents of mailbox ___ (mnemonic = LDA)"

Range("F25").Value = 0  'Set accumulator to zero

Set TempInput = Range(Chr$(CorrectColumn(Val(Address))) & CorrectRow(Val(Address)))  'Gather the contents of the memory cell

Accumulator = TempInput    'Set accumulator to contents of the memory cell

If Accumulator > 9999 Then Accumulator = 0    'Accumulator is restricted to numbers with four digits

If Accumulator < -9999 Then Accumulator = 0    'Accumulator is restricted to numbers with four digits

Range("F25").Value = Accumulator

If Accumulator < 0 Then Range("E25").Value = "-" Else Range("E25").Value = "+"    'Checks to see whether accumulator value is negative or positive (or zero); show + or - on-screen

End If
```

CARDIAC and LMC: the following routine adds values— either from the accumulator (or calculator) or to the accumulator (or calculator) (ADD):

```
If (Computer = 0 And OPCode = 2) Or (Computer = 1 And OPCode = 1) And LineofCode <> "" Then

If Computer = 0 Then Code = "Add contents of cell ___ into accumulator (mnemonic = ADD)"
```

```
If Computer = 1 Then Code = "Add contents of value in mailbox ___ to value in
calculator (mnemonic = ADD)"

Set TempAccum = Range("F25")  'Get what's in accumulator first

Set TempInput = Range(Chr$(CorrectColumn(Val(Address))) &
CorrectRow(Val(Address)))    'Gather the contents of the memory cell

Accumulator = TempAccum + TempInput        'Add the value in the memory cell
from the value in the accumulator

If Accumulator > 9999 Then Accumulator = 0      'Accumulator is restricted to
numbers with four digits

If Accumulator < -9999 Then Accumulator = 0     'Accumulator is restricted to
numbers with four digits

Range("F25").Value = Accumulator

If Accumulator < 0 Then Range("E25").Value = "-" Else Range("E25").Value =
"+"      'Checks to see whether accumulator value is negative or positive (or
zero); show + or - on-screen

End If
```

CARDIAC only: the following routine is a conditional jump (TAC):

```
If Computer = 0 And OPCode = 3 And LineofCode <> "" Then

Code = "Test accumulator contents: if the contents are negative, jump to cell
   ___; if the contents are positive (or zero; CARDIAC treats zero as positive)
then just move to the next cell (mnemonic = TAC)"

    If Accumulator < 0 Then

     'Find the correct row and correct column to jump to, first clearing the
     memory cell with a clear coat:

     Range(Chr$(Columnct - 1) & Rowct).Select
        With Selection.Interior
            .Pattern = xlNone
            .TintAndShade = 0
            .PatternTintAndShade = 0
        End With

     Rowct = CorrectRow(Val(Address) - 1)
     Columnct = CorrectColumn(Val(Address) - 1)

     ProgramStep = Val(Address) - 1

    End If
End If
```

LMC only: the following routine is a conditional jump (BRZ):

```
If Computer = 1 And OPCode = 7 And LineofCode <> "" Then

Code = "Test calculator contents: if the contents equal zero, jump to mailbox
   ___; anything else, then just move to the next cell (mnemonic = BRZ)"
```

```
    If Accumulator = 0 Then

    'Find the correct row and correct column to jump to, first clearing the
memory cell with a clear coat:

    Range(Chr$(Columnct - 1) & Rowct).Select
        With Selection.Interior
            .Pattern = xlNone
            .TintAndShade = 0
            .PatternTintAndShade = 0
        End With

    Rowct = CorrectRow(Val(Address) - 1)
    Columnct = CorrectColumn(Val(Address) - 1)

    ProgramStep = Val(Address) - 1

    End If

End If
```

LMC only: the following routine is another conditional jump (BRP):

```
If Computer = 1 And OPCode = 8 And LineofCode <> "" Then

Code = "Test calculator contents: if the contents are positive or zero, jump
to mailbox ___; anything else, then just move to the next cell (mnemonic =
BRP)"

    If Accumulator >= 0 Then

    'Find the correct row and correct column to jump to, first clearing the
memory cell with a clear coat:

    Range(Chr$(Columnct - 1) & Rowct).Select
        With Selection.Interior
            .Pattern = xlNone
            .TintAndShade = 0
            .PatternTintAndShade = 0
        End With

    Rowct = CorrectRow(Val(Address) - 1)
    Columnct = CorrectColumn(Val(Address) - 1)

    ProgramStep = Val(Address) - 1

    End If

End If
```

CARDIAC only: the following routine shifts digits in the accumulator left and right (SFT):

```
If Computer = 0 And OPCode = 4 And LineofCode <> "" Then

Code = "Shift accumulator x places left (x = first digit of operand) and y
places right (y = second digit of operand) (mnemonic = SFT)"


'Obtain the first and second digits of the operand:
```

```
If Val(Address) < 10 Then
    FirstDigit = 0
    SecondDigit = Val(Address)
End If

If Val(Address) >= 10 Then
    FirstDigit = Int(Val(Address) / 10)
    SecondDigit = Val(Address) - FirstDigit * 10
End If

'First, shift digits left "x" units:

If FirstDigit = 1 Then Accumulator = (Accumulator - (Int(Accumulator / 1000)) * 1000) * 10

If FirstDigit = 2 Then Accumulator = (Accumulator - (Int(Accumulator / 100)) * 100) * 100

If FirstDigit = 3 Then Accumulator = (Accumulator - (Int(Accumulator / 10)) * 10) * 1000

If FirstDigit > 3 Then Accumulator = 0        'Shifting by anything more than three to the left clears out the accumulator

'Then, shift digits right "y" units:

Accumulator = Int(Accumulator / (10 ^ SecondDigit))

'Then error check the Accumulator:

If Accumulator > 9999 Then Accumulator = 0        'Accumulator is restricted to numbers with four digits

If Accumulator < -9999 Then Accumulator = 0       'Accumulator is restricted to numbers with four digits

'And print out its new value:

Range("F25").Value = Accumulator

If Accumulator < 0 Then Range("E25").Value = "-" Else Range("E25").Value = "+"        'Checks to see whether accumulator value is negative or positive (or zero); show + or - on-screen

End If
```

CARDIAC only: the following routine prints the contents of a memory cell to the output (OUT):

```
If Computer = 0 And OPCode = 5 And LineofCode <> "" Then

Code = "Print contents of cell ___ on output card (mnemonic = OUT)"

Set TempInput = Range(Chr$(CorrectColumn(Val(Address))) & CorrectRow(Val(Address)))        'Gather the contents of the memory cell

Range("R" & OutputRow).Value = TempInput

OutputRow = OutputRow + 1        'Increment the output card

End If
```

CARDIAC and LMC: the following routine stores the value present in the accumulator (or calculator) (STO/STA):

```
If (Computer = 0 And OPCode = 6) Or (Computer = 1 And OPCode = 3) And
LineofCode <> "" Then

If Computer = 0 Then Code = "Store contents of accumulator into cell ___
(mnemonic = STO)"

If Computer = 1 Then Code = "Store contents of calculator into mailbox ___
(mnemonic = STA)"

Set TempAccum = Range("F25")    'Get what's in the accumulator

Range(Chr$(CorrectColumn(Val(Address))) & CorrectRow(Val(Address))).Value =
TempAccum 'Copy value in accumulator into the address listed

End If
```

LMC only: the following routine subtracts a number from the accumulator (or calculator) (SUB):

```
If (Computer = 0 And OPCode = 7) Or (Computer = 1 And OPCode = 2) And
LineofCode <> "" Then

If Computer = 0 Then Code = "Subtract contents of cell ___ into accumulator
(mnemonic = SUB)"

If Computer = 1 Then Code = "Subtract contents of value in mailbox ___ to
value in calculator (mnemonic = SUB)"

Set TempAccum = Range("F25")    'Get what's in accumulator first

Set TempInput = Range(Chr$(CorrectColumn(Val(Address)))
& CorrectRow(Val(Address)))    'Gather the contents of the memory cell

Accumulator = TempAccum - TempInput     'Subtract the value in the memory
cell from the value in the accumulator

If Accumulator > 9999 Then Accumulator = 0    'Accumulator is restricted to
numbers with four digits

If Accumulator < -9999 Then Accumulator = 0   'Accumulator is restricted to
numbers with four digits

Range("F25").Value = Accumulator

If Accumulator < 0 Then Range("E25").Value = "-" Else Range("E25").Value =
"+"     'Checks to see whether accumulator value is negative or positive (or
zero); show + or - on-screen

End If
```

CARDIAC and LMC: the following routine is an unconditional jump (JMP/BRA):

```
If (Computer = 0 And OPCode = 8) Or (Computer = 1 And OPCode = 6) And
LineofCode <> "" Then
```

```vb
If Computer = 0 Then Code = "Unconditional jump to instruction in cell ___ (mnemonic = JMP)"

If Computer = 1 Then Code = "Unconditional branch to instruction in mailbox ___ (mnemonic = BRA)"


'Records address in memory cell 99 before jumping (for CARDIAC only):

If Computer = 0 And ProgramStep < 99 Then Range("L19").Value = "8" & (ProgramStep + 1)

If Computer = 0 And ProgramStep >= 99 Then Range("L19").Value = "800"

If Accumulator <> 0 Then

'Find the correct row and correct column to unconditionally jump to, first clearing the memory cell with a clear coat:

 Range(Chr$(Columnct - 1) & Rowct).Select
   With Selection.Interior
        .Pattern = xlNone
        .TintAndShade = 0
        .PatternTintAndShade = 0
    End With

    Rowct = CorrectRow(Val(Address) - 1)
    Columnct = CorrectColumn(Val(Address) - 1)

    ProgramStep = Val(Address) - 1

End If

If Accumulator = 0 Then

'Find the correct row and correct column to unconditionally jump to, first clearing the memory cell with a clear coat:

 Range(Chr$(Columnct - 1) & Rowct).Select
   With Selection.Interior
        .Pattern = xlNone
        .TintAndShade = 0
        .PatternTintAndShade = 0
    End With

    Rowct = CorrectRow(Val(Address) - 1)
    Columnct = CorrectColumn(Val(Address) - 1)

    ProgramStep = Val(Address) - 1
End If

End If
```

CARDIAC and LMC: the following routine terminates the program (HRS/HLT or COB):

```vb
If (Computer = 0 And OPCode = 9) Or (Computer = 1 And OPCode = 0 And Range(Chr$(Columnct) & Rowct).Value <> "") And LineofCode <> "" Then

If Computer = 0 Then Code = "Halt machine and reset program counter to ___ (mnemonic = HRS)"
```

```
If Computer = 1 Then Code = "End the program (mnemonic = HLT or COB)"

Range("A25").Value = Val(Address)      'Sets the bug to address value

Range("F25").Value = 0       'Sets accumulator to zero

Exit Do       'Exit Do Loop/End Program

End If
```

If the debugger option is turned on, once a line of code is run, a dialog box pops up explaining the code.

```
'Debugger, Part 2: A post-run explanation of the line of code:

If Range("L30").Value = 1 Then MsgBox ("Just ran: LineofCode = " & LineofCode
& " and opcode is " & OPCode & " and Address is " & Address & ". Which means:
" & Code & ".")
```

Several error-checking routines follow.

```
'Create dummy OPCode if the cell is blank:

If Range(Chr$(Columnct) & Rowct).Value <> "" Then OPCode = -999       'Makes a
useless OPCode if the cell is blank

'Repaint current memory cell with a clear coat

  Range(Chr$(Columnct - 1) & Rowct).Select
    With Selection.Interior
        .Pattern = xlNone
        .TintAndShade = 0
        .PatternTintAndShade = 0
    End With
```

Here's code for the "slow down" option, if it's selected (by typing a 1 into cell L29):

```
'If Pause-between-steps is selected, pause:

If Range("L29").Value = 1 Then For Pause = 1 To 100000000: Next Pause
```

And here are more error-checking routines:

```
'Go to the next program step; if greater than 99, go back to 00

ProgramStep = ProgramStep + 1
If ProgramStep > 99 Then ProgramStep = 0


' Increment bug to next row (as long as there wasn't an unconditional or
conditional jump); if too high, reset row and move bug over to the next
column of memory cells:

If (Computer = 0 And OPCode <> 8) Or (Computer = 0 And (OPCode <> 3 And
Accumulator < 0)) Or (Computer = 1 And (OPCode <> 6 Or ((OPCode <> 7 Or
OPCode <> 8) And Accumulator < 0))) Then Rowct = Rowct + 1
```

```
If Rowct > 21 Then
    Rowct = 5
    Columnct = Columnct + 2
    If Columnct > 76 Then Columnct = 66
End If

If (Computer = 0 And OPCode <> 9) Or (Computer = 1 And OPCode <> 0) Then
Range("A25").Value = ProgramStep     'Sets the bug display to correct memory
cell (as long as the program hasn't halted)

Loop

End Sub
```

Two functions— for checking the correct row and column placement of the bug on the Excel worksheet— conclude the macro.

```
Function CorrectRow(Addr As Integer) As Integer

'Calculate which Excel row the associated Address refers to:

Dim Row As Integer

If Addr < 0 Then Row = 5
If Addr < 17 Then Row = Addr + 5
If Addr > 16 And Addr < 34 Then Row = Addr - 17 + 5
If Addr > 33 And Addr < 51 Then Row = Addr - 17 * 2 + 5
If Addr > 50 And Addr < 68 Then Row = Addr - 17 * 3 + 5
If Addr > 67 And Addr < 85 Then Row = Addr - 17 * 4 + 5
If Addr > 84 And Addr < 100 Then Row = Addr - 17 * 5 + 5

If Addr = 0 Or Addr = 17 Or Addr = 34 Or Addr = 51 Or Addr = 68 Or Addr = 85
Then Row = 5

CorrectRow = Row    'Returns correct row

End Function

Function CorrectColumn(Addr As Integer) As Integer

'Calculate which Excel column the associated Address refers to:

Dim Column As Integer

If Addr < 0 Then Column = 66
If Addr < 17 Then Column = 66
If Addr > 16 And Addr < 34 Then Column = 68
If Addr > 33 And Addr < 51 Then Column = 70
If Addr > 50 And Addr < 68 Then Column = 72
If Addr > 67 And Addr < 85 Then Column = 74
If Addr > 84 And Addr < 100 Then Column = 76

If Addr = 0 Then Column = 66
If Addr = 17 Then Column = 68
If Addr = 34 Then Column = 70
If Addr = 51 Then Column = 72
If Addr = 68 Then Column = 74
If Addr = 85 Then Column = 76

CorrectColumn = Column  'Returns correct column (the ASCII code)

End Function
```

several simple example programs

Our first example program is designed to run on the CARDIAC (make sure cell L28 is set to 0). You will need to input any three positive integers before running the program (type the numbers into cells P4, P5, and P6). Then type the program in, starting in address 010 (you can either type in the machine language codes or the mnemonics into the spreadsheet):*

CARDIAC Program No. 6-1: Adding and Subtracting

ADDRESS	CONTENTS	MNEMONIC	OPERAND	COMMENTS
10	020	INP	20	Read first input into cell 20.
11	021	INP	21	Read second input into cell 21.
12	022	INP	22	Read third input into cell 22.
13	120	CLA	20	Clear accumulator contents, replacing them with the contents of cell 20.
14	221	ADD	21	Add the contents of cell 21 to the value in the accumulator.
15	722	SUB	22	Subtract the contents of cell 22 from the value in the accumulator.
16	623	STO	23	Store the contents of the accumulator into cell 23.
17	523	OUT	23	Print the result of the calculation (which is found in cell 23).
18	900	HRS	00	Halt and reset.

Set the instructor register to address 10 (type 10 into cell A25), which will start the bug at address 10— the first "line" of the program. Click the "START" button to run the program. The first two input numbers are added together; the difference between that sum and the third inputted number is outputted to the screen.

Our second example program is designed to run on the LMC (make sure cell L28 is set to 1), and performs the same addition-subtraction routine as the CARDIAC program above. Again, you will need to input any three integers before running the program (type the numbers into cells P4, P5, and P6). Type the program in, starting in address 10 (like with the CARDIAC, you have two options: entering the code as machine language or as mnemonics):†

---

* If you choose to type in the mnemonics, you'll have to supply an address for each instruction as well. For instance, the first line of machine language code— 020— would need to be keyed in as INP 20 (INP into cell U13 and 20 into cell V13). In addition, once all the mnemonics and addresses have been entered into the appropriate locations of the spreadsheet, you'll need to type 1 inside cell L31 before clicking the "START" button, which will convert the mnemonics into machine language code that the macro will be able to run.

† If choosing to use mnemonics, though, you'll have to supply a mailbox (address) number for each instruction— except INP (fetches a value from the inbox, placing it into the calculator), OUT (takes the calculator's value and copies it to the outbox), and COB (ends the program).

LMC Program No. 6-1: Adding and Subtracting

MAILBOX	CONTENTS	MNEMONIC	OPERAND	COMMENTS
10	901	INP		Read the first input and copy its value into the calculator.
11	330	STA	30	Store the contents of the calculator into mailbox 30.
12	901	INP		Read the second input and copy its value into the calculator.
13	331	STA	31	Store the contents of the calculator into mailbox 31.
14	901	INP		Read the third input and copy its value into the calculator.
15	332	STA	32	Store the contents of the calculator into mailbox 31.
16	530	LDA	30	Load the value in mailbox 30 into the calculator.
17	131	ADD	31	Add the value in mailbox 31 into the value currently contained in the calculator.
18	232	SUB		Subtract the value in mailbox 31 from the value currently contained in the calculator.
19	902	OUT		Copy the value contained in the calculator to the outbox.
20	000	COB		End the program.

Set the program counter to mailbox 10 (type 10 into cell A25), which will start running the program at mailbox 10. Then click the "START" button to begin. If necessary, you can interrupt the macro by repeatedly pressing the Escape key.

section 7.

a cardiac-little man assembler

Juggling both opcodes and addresses for the CARDIAC and LMC isn't a particularly intuitive way to write programs. Using the mnemonics associated with the opcodes, along with allowing space for the declaration and use of variables (which are assigned addresses behind the scenes), would greatly ease the difficulties of programming these two papers computers. Essentially, we need to write an assembler for the CARDIAC and the LMC; as we did with the machine language emulator, let's package the assembler for both paper computers into one macro.

getting started
First, you'll need to make sure that you have the "Developer" tab in the Excel "Ribbon" (the strip of options right below the menu) available. Go to "File," then "Options," then "Customize Ribbon." Finally, check off the "Developer" option. Even if you don't have a "Ribbon," you'll need to make sure that the "Developer" option is selected in order to proceed.

Open up a new Excel spreadsheet, and then save it with the filename "Assembler for CARDIAC-LMC." Make sure your workbook is "Macro-Enabled" when saving.

setting up the worksheet
First, right click the worksheet's tab name, type "CARDIAC-LMC," and then click back inside the worksheet.

In cell J7, type 0. The format of the 0, however, needs leading zeros, so it looks like this: 0000. In order to make that change, right click the cell and select "Format Cells." Once there, select the "Custom" option and, in the text box labeled "Type:" type in 0000 and press the "OK" button. Also bold cells J7 and I7, and right-justify I7. In cell I7, type "Current Value in Accumulator:".

In cell E1, type "Note: Start your assembly program on line 1. Column B is for mnemonics; Column C is for variables, numerical quantities, or numerical commands;" and in cell E2 type "Column D is for an initial value assigned to a variable (so only the mnemonic DAT, along with a variable name, can be in columns B and C on that line)."

Type 0 into cell J4, and 1 into cell J5; insert borders around both cells. Next, right-justify cells I4 and I5. In cell I4, type "Use which computer? (CARDIAC = 0, LMC = 1)" and in cell I5, type "See Steps? (0 = No, 1 = Yes)." In cell E1, type "Note: Start your assembly program on line 1 (next to 'start'). Column A is for labels or variable names; Column B is for mnemonics;" and in cell E2, type "Column C is for variables, numerical quantities, or numerical commands. For initializing variables, use the mnemonic DAT." (In other words, Column A is for the labels, Column B is for the opcodes, and Column C is for the operands.)

Type the word "start" into cell A1. Finally, in cell A99 type 99, in cell B99 type "JMP," and in cell C99 type 00.

Your worksheet should look like the following:

coding the macro

We will first need to insert a button that will run our macro. Click on the "Developer" tab, select the "Insert" dropdown, and click the gray button. Draw a rectangular button, starting around cell G10; once you let go of the mouse button, a dialog box will pop up asking you which macro you'd like to assign to the button. Click "Cancel" for now, and instead right click the button and select "Edit Text." Type "RUN PROGRAM" and click somewhere outside of the button. Your worksheet should resemble this screenshot:

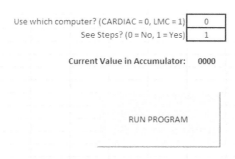

It's now time to insert the macro. Right click the "START" button, select "Assign Macro," and click the "New" button. Underneath the subroutine called `Sub Button1_Click()`, type the following:

```
'Run Program macro

Dim InputValue As Variant
Dim Accumulator As Long
Dim Rowct As Integer
Dim Computer As Integer
Dim CommandValue As Variant
Dim VariableName(1 To 100) As String
Dim VariableValue(1 To 100) As Integer
Dim LabelName(1 To 100) As String
Dim VariableIndex As Integer
Dim Mnemonic As String
Dim DoesVariableExist As Integer
Dim FirstDigit As Integer
Dim SecondDigit As Integer
Dim IsIttheEnd As Integer
```

The `Dim` statements above declare a number of important variables for later use. Some variables are of type `String`, meaning that characters of any sort can be stored in them, while others are of type `Integer`, which is self-explanatory. The `Long` variable allows for a more precise number than the `Integer` type, while the `Variant` type can handle a variety of inputs. In addition, there are three arrays: `VariableName`, `VariableValue`, and `LabelName`. Arrays allow us to organize and index multiple inputs, all of the same type; here, specifically, they will keep track of the different variables and labels declared, as well as their assigned values.

Sprinkled throughout the code are comments— shown as phrases that begin with apostrophes— that have been added for clarity.

Throughout the rest of the macro, we will have to keep alternating between CARDIAC code and LMC code— the variable `Computer` will keep track: 0 = CARDIAC, 1 = LMC— as shown below:

```
Computer = Range("J4").Value        'Tells macro which computer we're using--
CARDIAC (value = 0) or LMC (value = 1)
```

Next, we will assign several variables initial values, and set up the workspace for the CARDIAC or the LMC appropriately:

```
'Initialize important variables

Rowct = 1
Accumulator = 0
VariableIndex = 1
IsIttheEnd = 0


'Get line 99 ready (for the CARDIAC or the LMC):

If Computer = 0 Then
    Range("A99").Value = 99
    Range("B99").Value = "JMP"
    Range("C99").Value = 0
End If
```

```
If Computer = 1 Then
    Range("A99").Value = ""
    Range("B99").Value = ""
    Range("C99").Value = ""
End If
```

If there are any variables declared in the assembly program using the mnemonic DAT— which is canonical for the LMC but non-canonical for the CARDIAC (but utilized here regardless)— the macro stores them into the arrays; in addition, any program labels are stored into an array as well:

```
'Check to see if any "DAT" statements are used (for either CARDIAC or LMC);
if so, set the variable name into the array and assign the initial value (if
any) next to the variable name:

For X = 1 To 98
    If Range("B" & X).Value = "DAT" Then
        VariableName(X) = Range("A" & X)
        VariableValue(X) = Range("C" & X)
    End If
Next X


'Check to see if there are any labels:

For X = 1 To 98
    If Range("A" & X).Value <> "DAT" And Range("A" & X).Value <> "" Then
        LabelName(X) = Range("A" & X)
    End If
Next X
```

Next begins a `While...Wend` loop to decode and run the assembly program. Inside the loop is the code for the debugger, all the mnemonics for the CARDIAC and LMC, and some other housekeeping routines:

```
'The loop to run a CADIAC/LMC program starts here:

While IsIttheEnd = 0              'Until the variable "IsIttheEnd" equals 1, the
loop continues....

'Debugger: about to run a line of code:

If Range("J5").Value = 1 Then MsgBox ("About to run line " & Rowct & ": " &
Range("B" & Rowct) & " " & Range("C" & Rowct))


'Get mnemonic

Mnemonic = Range("B" & Rowct)


'Get variable name or value: check to see which it is first, then assign
appropriately:

CommandValue = Range("C" & Rowct)
```

```
If Not IsNumeric(CommandValue) Then

    'First, check to see if the variable has been used before--first as a
    variable, then as a label:

    DoesVariableExist = 0

        For X = 1 To 100

            If VariableName(X) = CommandValue Then
                DoesVariableExist = 1
                VariableIndex = X
            End If

        Next X

        For X = 1 To 100

            If LabelName(X) = CommandValue Then
                DoesVariableExist = 1
                VariableIndex = X
            End If

        Next X

    'Only if variable hasn't been used yet should it be assigned to a new
    spot in the VariableName array:

    If DoesVariableExist = 0 Then

        VariableIndex = 1

        While VariableName(VariableIndex) <> ""

            VariableIndex = VariableIndex + 1

        Wend

    'Deposits the variable name into the next available array index:

        VariableName(VariableIndex) = CommandValue

        End If

End If


'CARDIAC: Input in a value into a variable:

If Computer = 0 And Mnemonic = "INP" Then

    InputValue = InputBox("Input the value for variable " & CommandValue &
    ": ", "INPUT", 1)

    VariableValue(VariableIndex) = InputValue

    Rowct = Rowct + 1

End If
```

```
'LMC: Input in a new value into the accumulator:

If Computer = 1 And Mnemonic = "INP" Then

    InputValue = InputBox("Input in new value for the accumulator: ",
    "INPUT", 1)

    Accumulator = InputValue

    If Accumulator > 9999 Then Accumulator = 0      'Accumulator is
    restricted to numbers with four digits

    If Accumulator < -9999 Then Accumulator = 0     'Accumulator is
    restricted to numbers with four digits

    'Print out accumulator's new value:

    Range("J7").Value = Accumulator

    Rowct = Rowct + 1

End If


'CARDIAC or LMC: Clear the accumulator and insert contents of a
number/variable (CARDIAC); load a value from number/variable and insert into
accumulator (LMC):

If (Computer = 0 And Mnemonic = "CLA") Or (Computer = 1 And Mnemonic = "LDA")
Then

    If Not IsNumeric(CommandValue) Then Accumulator =
    VariableValue(VariableIndex) Else Accumulator = Val(CommandValue)

    If Accumulator > 9999 Then Accumulator = 0      'Accumulator is
    restricted to numbers with four digits

    If Accumulator < -9999 Then Accumulator = 0     'Accumulator is
    restricted to numbers with four digits

    Rowct = Rowct + 1

End If


'CARDIAC and LMC: Add the value of the variable/number to the value in the
accumulator:

If Mnemonic = "ADD" Then

    If Not IsNumeric(CommandValue) Then Accumulator = Accumulator +
    VariableValue(VariableIndex) Else Accumulator = Accumulator +
    Val(CommandValue)

    If Accumulator > 9999 Then Accumulator = 0      'Accumulator is
    restricted to numbers with four digits

    If Accumulator < -9999 Then Accumulator = 0     'Accumulator is
    restricted to numbers with four digits

    Rowct = Rowct + 1

End If
```

```
'CARDIAC and LMC: Subtract the contents of the variable/number from the value
in the accumulator:

If Mnemonic = "SUB" Then

    If Not IsNumeric(CommandValue) Then Accumulator = Accumulator -
    VariableValue(VariableIndex) Else Accumulator = Accumulator -
    Val(CommandValue)

    If Accumulator > 9999 Then Accumulator = 0      'Accumulator is
    restricted to numbers with four digits

    If Accumulator < -9999 Then Accumulator = 0     'Accumulator is
    restricted to numbers with four digits

    Rowct = Rowct + 1

End If


'CARDIAC and LMC: Store the contents of the accumulator into a variable:

If (Computer = 0 And Mnemonic = "STO") Or (Computer = 1 And Mnemonic = "STA")
Then

    VariableValue(VariableIndex) = Accumulator

    Rowct = Rowct + 1

End If


'CARDIAC: Display the output on-screen (first determine if the output is the
value of a variable, or simply a number):

If Computer = 0 And Mnemonic = "OUT" Then

    If Not IsNumeric(CommandValue) Then MsgBox ("The output is: " &
    VariableValue(VariableIndex))

    If IsNumeric(CommandValue) Then MsgBox ("The output is: " &
    Val(CommandValue))

    Rowct = Rowct + 1

End If


'LMC: Output the value showing on the accumulator:

If Computer = 1 And Mnemonic = "OUT" Then

    MsgBox ("The output of the accumulator is: " & Accumulator)

    Rowct = Rowct + 1

End If


'CARDIAC and LMC: Halt the machine:

If (Computer = 0 And Mnemonic = "HRS") Or (Computer = 1 And (Mnemonic = "HLT"
Or Mnemonic = "COB")) Then
```

```
        IsIttheEnd = 1

        Rowct = Rowct + 1

End If


'CARDIAC and LMC: Unconditional jump, while "remembering" the current line of
code in line 99 (CARDIAC); unconditional jump (LMC):

If (Computer = 0 And Mnemonic = "JMP") Or (Computer = 1 And Mnemonic = "BRA")
Then

    If Computer = 0 Then Range("C99").Value = Rowct

    If IsNumeric(CommandValue) Then Rowct = Val(CommandValue)

    If Not IsNumeric(CommandValue) Then
        For X = 1 To 98
            If LabelName(X) = CommandValue Then Rowct = X
        Next X
    End If

End If


'CARDIAC: Conditional jump--check to see if accumulator is negative first

If Computer = 0 And Mnemonic = "TAC" Then

    If (IsNumeric(CommandValue) And Accumulator < 0) Then
    Rowct = Val(CommandValue) Else Rowct = Rowct + 1

    If (Not IsNumeric(CommandValue) And Accumulator < 0) Then
        For X = 1 To 98
            If LabelName(X) = CommandValue Then Rowct = X
        Next X
    End If

End If


'LMC: Conditional jump--check to see if accumulator is zero to jump

If Computer = 1 And Mnemonic = "BRZ" Then

    If (IsNumeric(CommandValue) And Accumulator = 0) Then
    Rowct = Val(CommandValue) Else Rowct = Rowct + 1

    If (Not IsNumeric(CommandValue) And Accumulator = 0) Then
        For X = 1 To 98
            If LabelName(X) = CommandValue Then Rowct = X
        Next X
    End If

End If


'LMC: Conditional jump--check to see if accumulator is zero or positive to
jump

If Computer = 1 And Mnemonic = "BRP" Then
```

```
    If (IsNumeric(CommandValue) And Accumulator >= 0) Then
    Rowct = Val(CommandValue) Else Rowct = Rowct + 1

    If (Not IsNumeric(CommandValue) And Accumulator >= 0) Then
        For X = 1 To 98
            If LabelName(X) = CommandValue Then Rowct = X
        Next X
    End If

End If

'CARDIAC: Shift accumulator x places left (x = first digit of address) and y
places right (y = second digit of address)

If Computer = 0 And Mnemonic = "SFT" Then

'Obtain the first and second digits of the address:

FirstDigit = 0
SecondDigit = 0

If Val(CommandValue) < 10 Then
    FirstDigit = 0
    SecondDigit = Val(CommandValue)
End If

If Val(CommandValue) >= 10 Then
    FirstDigit = Int(Val(CommandValue) / 10)
    SecondDigit = Val(CommandValue) - FirstDigit * 10
End If

'First, shift digits left "x" units:

If FirstDigit = 1 Then Accumulator = (Accumulator - (Int(Accumulator / 1000))
* 1000) * 10

If FirstDigit = 2 Then Accumulator = (Accumulator - (Int(Accumulator / 100))
* 100) * 100

If FirstDigit = 3 Then Accumulator = (Accumulator - (Int(Accumulator / 10)) *
10) * 1000

If FirstDigit > 3 Then Accumulator = 0       'Shifting by anything more than
three to the left clears out the accumulator

'Then, shift digits right "y" units:

Accumulator = Int(Accumulator / (10 ^ SecondDigit))

'Then error check the Accumulator:

If Accumulator > 9999 Then Accumulator = 0        'Accumulator is restricted
to numbers with four digits

If Accumulator < -9999 Then Accumulator = 0       'Accumulator is restricted to
numbers with four digits

'And print out accumulator's new value:

Range("J7").Value = Accumulator

Rowct = Rowct + 1

End If
```

```
'Refresh accumulator display:

Range("J7").Value = Accumulator

Wend

End Sub
```

several simple example programs

Our first example is designed to run on the CARDIAC assembler (make sure cell J4 is set to 0). You will input two numbers, one at a time, when input boxes pop up. Notice that, by convention, the mnemonics are uppercase, while the labels, operations, operands, and variables are lowercase.

CARDIAC Program No. 7-1: Adding and Subtracting

LABEL	MNEMONIC	OPERAND	COMMENTS
start	INP	first	Asks the user for the first input; stores the input into the variable "first."
	INP	second	Asks the user for the second input; stores the input into the variable "second."
	CLA	first	Clears the accumulator's contents, replacing them with the value of the variable "first."
	ADD	second	Adds the contents of the variable "second" to the contents in the accumulator.
	SUB	third	Subtracts the contents of the accumulator by the value of the variable "third."
	STO	result	Stores the contents of the accumulator into the variable "result."
	OUT	result	Prints the value of the variable "result."
	HRS		Ends the program.
third	DAT	10	Assigns the variable "third" an initial value of 10.

Once the code is typed in, click the "RUN PROGRAM" button to run. The program adds the two inputs together, subtracts 10, and then outputs the result.

Our second example is designed to run on the LMC assembler (make sure cell J4 is set to 1). You will input two numbers, one at a time, when input boxes pop up. Again, notice that the mnemonics are uppercase, while the labels, operands, and variables are lowercase.

LMC Program No. 7-1: Adding and Subtracting

LABEL	MNEMONIC	OPERAND	COMMENTS
start	INP		Asks the user for the first input; copies the input into the calculator.
	STA	first	Stores the contents of the calculator into the variable "first."

	INP		Asks the user for the second input; copies the input into the calculator.
	STA	second	Stores the contents of the calculator into the variable "second."
	LDA	first	Loads the value of the variable "first" into the calculator.
	ADD	second	Adds the contents of the variable "second" to the contents in the calculator.
	SUB	third	Subtracts the contents of the calculator by the value of the variable "third."
	OUT		Copy the value contained in the calculator to the outbox.
	COB		Ends the program.
third	DAT	10	Assigns the variable "third" an initial value of 10.

Once the code is typed in, click the "RUN PROGRAM" button to run. Just like the analogous CARDIAC program, this LMC program adds the two inputs together, subtracts 10, and then outputs the result. If necessary, you can interrupt the macro by repeatedly pressing the Escape key.

section

an instructo emulator and compiler

The Instructo has a vast instruction set; procure a rare original model and you'll find running a program to be taxing on two levels: decoding each mnemonic and moving strips of paper in and out of the cardboard.

We can, at the very least, eliminate the physical tedium of running an Instructo program by using an Excel macro— well, actually, three macros. Although there are quite a few CARDIAC and LMC simulators available online, the emulator detailed in this section might very well be the only Instructo emulator ever made.

Our emulator will only be able to run machine language programs directly; but we will code a complier as well to translate assembly language instructions into machine language.

getting started

There are a number of versions of Excel available, so these macro-creation instructions will be sufficiently general.

First, you'll need to make sure that you have the "Developer" tab in the Excel "Ribbon" (the strip of options right below the menu) available. Go to "File," then "Options," then "Customize Ribbon." Finally, check off the "Developer" option. Even if you don't have a "Ribbon," you'll need to make sure that the "Developer" option is selected in order to proceed.

Open up a new Excel spreadsheet, and then save it with the filename "INSTRUCTO." Make sure your workbook is "Macro-Enabled" when saving.

setting up the worksheet

First, right click the worksheet's tab name, type "INSTRUCTO," and then click back inside the worksheet.

There's a lot to set up before we can code the macros. In cell B2, type "Program Step:" and in cell B4, type "Index Counter:". In cell F2, type "Input A:" and in cell F4, type "Input B (six inputs; use top one first):". In cell J2, type "Register A:" and in cell J4, type "Register B:". In cell M2, type "Output A" and in cell O2, type "Output B"; make sure to bold cells M2 and O2.

In cell B9, type "Reg.A" and in cell B10, type "Reg.B". Type in equals signs into cells C9 and C10, and then type "SS" into cells D9 and D10.

Type "A," "B," and "C" into cells B15, B16, and B17, respectively. Type the number 0 into cells C15, C16, and C17. Type "Line #" into cell E12, "Command" into cell F12, "(use 100 for IN)" into cell G12, and "Input" into cell H12; underline cells E12, F12, G12, and H12. In cell G11, type "SS or IN" and in cell H11, type "Special." In cell E113, type "100 (Index Counter)." And in cell F113, type:

```
=C4
```

Continuing, in cell J7, type "Pauses? (0 - No, 1 - Yes)" and in cell J8, type "See Steps? (0 - No, 1 - Yes)." Type 0 in cell K7, and 1 in cell K8.

Bold cell B7, and, in it, type "Compare Unit." Highlight cells B7 to D7, right click the area, and select "Format Cells." Under the "Alignment" tab, click the "Merge Cells" checkbox.

Bold cell B13. In cell B13, type "Jump Switches." Merge cells B13 and C13.

Bold cell E10 and, in it, type "Main Program." Merge cells E10 to I10.

Bold cell J10. In it, type "Main Storage." Merge cells J10 and K10.

Next, draw a border around the input/output "cards" by highlighting cells M2 to O13, right clicking the shaded region, selecting the "Format Cells" option again, clicking the "Border" tab, and then clicking the "Outline" button.

More borders need to be drawn, but only around these individual cells: G2, G4, G5, G6, G7, G8, G9, K7, and K8.

Now type 0 into cell E13. This zero needs to be two digits in length— i.e., 00. In order to make that change, right click the cell and select "Format Cells." Once there, select the "Custom" option and, in the text box labeled "Type:" type in 00 and press the "OK" button. Copy the formatting of this cell downward until reaching cell E112. Then populate the cells E14 to E112 with the numbers 01 to 99 (one number per cell).

The cells C2, C4, K2, K4, and F113 should also have the 00 format.

Place a border around cell J13, and enter in the following formula:

```
=E103
```

Place a border around cell K13, and enter in the following formula:

```
=IF(F103="","",F103)
```

Highlight cells J13 and K13, left click on the lower right corner of cell K13, and— which holding down the left mouse button— drag the mouse downward until reaching row 22. Then release the left mouse button.

Place your cursor in cell A1. The two screenshots below detail how your nearly finished Instructo workspace should appear.

First, the left side of your screen:

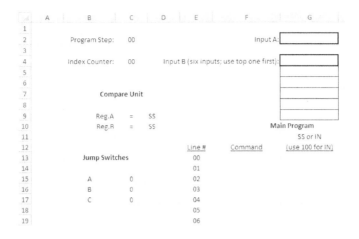

And next, the right side of your screen:

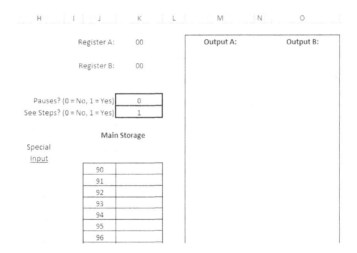

But we're not quite done, because we need to set up a workspace for an Instructo compiler: a program that will take an assembly language program and translate it into machine language.

Move your cursor over to cell Q1 and type "start." Next, starting in cell U1, type the instructions matching the screenshot below:

U	V	W	X	Y	Z	AA	AB	AC	AD	AE	AF	AG	AH	AI

Note: Start your assembly program on line 1 (next to "start"). Column Q is for labels or variable names; Column R is for mnemonics; Column S is for variables, numerical quantities, numerical commands, or IN (an index counter call). For initializing variables, use the mnemonic DAT. Leave no completely blank lines in your program; all DAT mnemonics must be at the end of the program.
You are allowed a maximum of ten read-write variables; each read-write variable declared thereafter is overwritten on top of the oldest ones.
Except with a DAT instruction, absolutely NO numeric operands are allowed here.

And now we're done.

coding the macros

We need to insert three buttons for our three macros: one to run a machine language program, one to clear the workspace, and one to compile an assembly program.

Click on the "Developer" tab, select the "Insert" dropdown, and click the gray button. Draw a square button, starting at around cell B19, which is about a cell in length and three cells in height; once you let go of the mouse button, a dialog box will pop up asking you which macro you'd like to assign to the button. Click "Cancel" for now, and instead right click the button and select "Edit Text." Type "START/STOP" and click somewhere outside of the button.

Draw a second button, adjacent to the first, starting at around cell C19. Then, change the text inside the button to read "RESET/CLEAR."

The workspace around the first two buttons on your worksheet should now resemble this screenshot:

Finally, go to cell U6 and draw a third small button. Edit the text in the button to read "COMPILE PROGRAM."

The space around this third button on your worksheet should look like this screenshot:

There is a lot of Visual Basic coding to do. Let's start by inserting the Reset/Clear macro. Right click the "RESET/CLEAR" button, select "Assign Macro," and click the "New" button. Underneath the subroutine that is labeled `Sub Button1_Click()`, type the following:

```
'Reset/Clear Macro

'Clear all active inputs/outputs:

    Range("C2").Select
    ActiveCell.FormulaR1C1 = "0"
    Range("C4").Select
    ActiveCell.FormulaR1C1 = "0"
```

```
    Range("C9").Select
    ActiveCell.FormulaR1C1 = "="
    Range("C10").Select
    ActiveCell.FormulaR1C1 = "="
    Range("C15").Select
    ActiveCell.FormulaR1C1 = "0"
    Range("C16").Select
    ActiveCell.FormulaR1C1 = "0"
    Range("C17").Select
    ActiveCell.FormulaR1C1 = "0"
    Range("G2:G6").Select
    Selection.ClearContents
    Selection.ClearContents
    Range("K2").Select
    ActiveCell.FormulaR1C1 = "0"
    Range("K4").Select
    ActiveCell.FormulaR1C1 = "0"
    Range("M3:O5000").Select
    Selection.ClearContents
    Selection.ClearContents
    Selection.ClearContents
    Selection.ClearContents
    Selection.ClearContents
    Range("K25").Select
    ActiveWindow.SmallScroll Down:=93
    Range("F103:F112").Select
    Selection.ClearContents
    Selection.ClearContents
    Selection.ClearContents
    Selection.ClearContents
    Selection.ClearContents
    Range("A1").Select
    Range("G5:G9").Select
    Selection.ClearContents
    Range("A1").Select

    'Clear any highlighted lines of code:

    Range("E13:E112").Select

    With Selection.Interior
        .Pattern = xlNone
        .TintAndShade = 0
        .PatternTintAndShade = 0
    End With

    'Move cursor to top left cell:

    Range("A1").Select
End Sub
```

Take note of the coded comments above: comments start with the apostrophe and bring greater clarity to the code.

Let's now insert the Stop/Start macro. Right click the "START/STOP" button, select "Assign Macro," and click the "New" button. Underneath the subroutine called `Sub Button2_Click()`, type the following:

```
' Run_Program Macro
```

The `Dim` statements below are used to declare some of the most important variables of the program; underneath these `Dim` statements, several of the variables are assigned initial values.

```
Dim Rowct As Integer
Dim OutputRowA As Integer
Dim OutputRowB As Integer
Dim InputBRow As Integer
Dim Code As String

OutputRowA = 3
OutputRowB = 3
InputBRow = 4

Set ProgramStep = Range("C2")     'Makes a variable called "ProgramStep" to be
the line number of the program

Rowct = 13 + ProgramStep          'Sets a variable named "Rowct" to be the
line number the program is on
```

Next, a `Do...While Loop` is established to run Instructo programs:

```
Do While StartCell <> "STOP"

Set StartCell = Range("F" & Rowct)   'Reads in a line of the code...
Set SS = Range("G" & Rowct)          'Reads the SS on that same line of code...
```

While the program is running, the line number corresponding to the program instruction is highlighted in yellow:

```
'Light up current line of code (in yellow)

 Range("E" & Rowct).Select
    With Selection.Interior
        .Pattern = xlSolid
        .PatternColorIndex = xlAutomatic
        .Color = 65535
        .TintAndShade = 0
        .PatternTintAndShade = 0
    End With
```

The emulator has a dedicated debugger. If the debugger is turned on (the option is selected by typing 1 into cell K8), then before each line of code is run, an anticipatory dialog box appears.

```
'Debugger, Part 1: Next line is the pre-run line of code:

If Range("K8").Value = 1 Then MsgBox ("About to run: LINE " & ProgramStep &
": " & Range("F" & Rowct) & ", " & Range("G" & Rowct))
```

The following routines print the data in the storage location to either Output A or B (PROA/PROB):

```
If StartCell = "PROA" Then

Code = "Print the data in storage location SS to Output A"

    Set TempOutput = Range("F" & (SS + 13))
        Range("M" & OutputRowA).Value = TempOutput
        OutputRowA = OutputRowA + 1  'Increment the output line # by 1

    ProgramStep = ProgramStep + 1  'Increment the line # of the program by 1
End If


If StartCell = "PROB" Then

Code = "Print the data in storage location SS to Output B"

    Set TempOutput = Range("F" & (SS + 13))
        Range("O" & OutputRowB).Value = TempOutput
        OutputRowB = OutputRowB + 1  'Increment the output line # by 1

    ProgramStep = ProgramStep + 1  'Increment the line # of the program by 1
End If
```

The following routines copy the data in the storage location to either Register A or B (LDRA/LDRB):

```
If StartCell = "LDRA" Then

Code = "Send the data located in storage location SS to Register A"

    Set TempRegister = Range("F" & (SS + 13))    'Set register to a new value
        Range("K2").Value = TempRegister

    ProgramStep = ProgramStep + 1

End If


If StartCell = "LDRB" Then

Code = "Send the data located in storage location SS to Register B"

    Set TempRegister = Range("F" & (SS + 13))    'Set register to a new value
        Range("K4").Value = TempRegister

    ProgramStep = ProgramStep + 1

End If
```

The following routines copy the data in Register A or B to the storage location (STRA/STRB):

```
If StartCell = "STRA" Then

Code = "Send the data in Register A to location SS"

Set TempRegister = Range("K2")
```

```
                Range("F" & (SS + 13)).Value = TempRegister    'Send value of
                register to storage

        ProgramStep = ProgramStep + 1

End If


If StartCell = "STRB" Then

Code = "Send the data in Register B to location SS"

Set TempRegister = Range("K4")

                Range("F" & (SS + 13)).Value = TempRegister    'Send value of
                register to storage

        ProgramStep = ProgramStep + 1

End If
```

The following routines sum the numbers in the storage location and Register A or B, copying the result into Register A or B (ADDA/ADDB):

```
If StartCell = "ADDA" Then

Code = "Sum the numbers in location SS and Register A, placing the result
into Register A"

Set TempRegister1 = Range("K2")
Set TempRegister2 = Range("F" & (SS + 13))
TempRegisterSum = TempRegister1 + TempRegister2

                Range("K2").Value = TempRegisterSum    'Send value of register
                to storage

        ProgramStep = ProgramStep + 1

End If


If StartCell = "ADDB" Then

Code = "Sum the numbers in location SS and Register B, placing the result
into Register B"

Set TempRegister1 = Range("K4")
Set TempRegister2 = Range("F" & (SS + 13))
TempRegisterSum = TempRegister1 + TempRegister2

                Range("K4").Value = TempRegisterSum    'Send value of register
                to storage

        ProgramStep = ProgramStep + 1

End If
```

The following routines test and report the sign of the number in Register A or B (CPRA/CPRB):

```
If StartCell = "CPRA" Then

Code = "Determine if the number in Register A is >, <, or = to the number
location SS; set the result into the Register A Compare Unit"

Set RegisterA = Range("K2")
Set Value = Range("F" & (SS + 13))            'Compare the register to
value in storage

If RegisterA > Value Then Range("C9").Value = ">"
If RegisterA = Value Then Range("C9").Value = "="
If RegisterA < Value Then Range("C9").Value = "<"

    ProgramStep = ProgramStep + 1

End If

If StartCell = "CPRB" Then

Code = "Determine if the number in Register B is >, <, or = to the number
location SS; set the result into the Register B Compare Unit"

Set RegisterB = Range("K4")
Set Value = Range("F" & (SS + 13))            'Compare the register to
value in storage

If RegisterB > Value Then Range("C10").Value = ">"
If RegisterB = Value Then Range("C10").Value = "="
If RegisterB < Value Then Range("C10").Value = "<"

    ProgramStep = ProgramStep + 1

End If
```

The following routines set the values of the Jump Switches:

```
If StartCell = "SJA0" Then

Code = "Set Jump Switch A to 0"

Range("C15").Value = 0

End If


If StartCell = "SJA1" Then

Code = "Set Jump Switch A to 1"

Range("C15").Value = 1

End If


If StartCell = "SJB0" Then

Code = "Set Jump Switch B to 0"
```

```
Range("C16").Value = 0

End If


If StartCell = "SJB1" Then

Code = "Set Jump Switch B to 1"

Range("C16").Value = 1

End If


If StartCell = "SJC0" Then

Code = "Set Jump Switch C to 0"

Range("C17").Value = 0

End If


If StartCell = "SJC1" Then

Code = "Set Jump Switch C to 1"

Range("C17").Value = 1

End If
```

The following routines perform conditional jumps:

```
If StartCell = "JALT" Then

Code = "Conditional jump: If Register A Compare Unit shows <, jump to SS"

Set ValueJ = SS

If Range("C9").Value = "<" Then ProgramStep = ValueJ Else ProgramStep = ProgramStep + 1

End If


If StartCell = "JBLT" Then

Code = "Conditional jump: If Register B Compare Unit shows <, jump to SS"

Set ValueJ = SS

If Range("C10").Value = "<" Then ProgramStep = ValueJ Else ProgramStep = ProgramStep + 1

End If


If StartCell = "JAGT" Then

Code = "Conditional jump: If Register A Compare Unit shows >, jump to SS"

Set ValueJ = SS
```

```
If Range("C9").Value = ">" Then ProgramStep = ValueJ Else ProgramStep =
ProgramStep + 1

End If


If StartCell = "JBGT" Then

Code = "Conditional jump: If Register B Compare Unit shows >, jump to SS"

Set ValueJ = SS

If Range("C10").Value = ">" Then ProgramStep = ValueJ Else ProgramStep =
ProgramStep + 1

End If


If StartCell = "JAEQ" Then

Code = "Conditional jump: If Register A Compare Unit shows =, jump to SS"

Set ValueJ = SS

If Range("C9").Value = "=" Then ProgramStep = ValueJ Else ProgramStep =
ProgramStep + 1

End If


If StartCell = "JBEQ" Then

Code = "Conditional jump: If Register B Compare Unit shows =, jump to SS"

Set ValueJ = SS

If Range("C10").Value = "=" Then ProgramStep = ValueJ Else ProgramStep =
ProgramStep + 1

End If


If StartCell = "JANL" Then

Code = "Conditional jump: If Register A Compare Unit doesn't show <, jump to SS"

Set ValueJ = SS

If Range("C9").Value = ">" Or Range("C9").Value = "=" Then ProgramStep =
ValueJ Else ProgramStep = ProgramStep + 1

End If


If StartCell = "JBNL" Then

Code = "Conditional jump: If Register B Compare Unit doesn't show <, jump to SS"

Set ValueJ = SS
```

```
    If Range("C10").Value = ">" Or Range("C10").Value = "=" Then ProgramStep =
    ValueJ Else ProgramStep = ProgramStep + 1

End If


If StartCell = "JANE" Then

Code = "Conditional jump: If Register A Compare Unit doesn't show =, jump to
SS"

Set ValueJ = SS

If Range("C9").Value = ">" Or Range("C9").Value = "<" Then ProgramStep =
ValueJ Else ProgramStep = ProgramStep + 1

End If


If StartCell = "JBNE" Then

Code = "Conditional jump: If Register B Compare Unit doesn't show =, jump to
SS"

Set ValueJ = SS

If Range("C10").Value = ">" Or Range("C10").Value = "<" Then ProgramStep =
ValueJ Else ProgramStep = ProgramStep + 1

End If


If StartCell = "JANG" Then

Code = "Conditional jump: If Register A Compare Unit doesn't show >, jump to
SS"

Set ValueJ = SS

If Range("C9").Value = "=" Or Range("C9").Value = "<" Then ProgramStep =
ValueJ Else ProgramStep = ProgramStep + 1

End If


If StartCell = "JBNG" Then

Code = "Conditional jump: If Register B Compare Unit doesn't show >, jump to
SS"

Set ValueJ = SS

If Range("C10").Value = "=" Or Range("C10").Value = "<" Then ProgramStep =
ValueJ Else ProgramStep = ProgramStep + 1

End If


If StartCell = "JIBD" Then

Code = "If there is any data in Input B, jump to location SS"
```

```
    If Range("G" & InputBRow).Value <> "" Then ProgramStep = SS Else
ProgramStep = ProgramStep + 1         'Jump to new line if there's
something showing in Input B

End If


If StartCell = "JJA0" Then

Code = "If Jump Switch A is set to 0, jump to location SS"

   If Range("C15").Value = 0 Then ProgramStep = SS          'Jump to new line
End If


If StartCell = "JJA1" Then

Code = "If Jump Switch A is set to 1, jump to location SS"

   If Range("C15").Value = 1 Then ProgramStep = SS          'Jump to new line
End If


If StartCell = "JJB0" Then

Code = "If Jump Switch B is set to 0, jump to location SS"

   If Range("C16").Value = 0 Then ProgramStep = SS          'Jump to new line
End If


If StartCell = "JJB1" Then

Code = "If Jump Switch B is set to 1, jump to location SS"

   If Range("C16").Value = 1 Then ProgramStep = SS          'Jump to new line
End If


If StartCell = "JJC0" Then

Code = "If Jump Switch C is set to 0, jump to location SS"

   If Range("C17").Value = 0 Then ProgramStep = SS          'Jump to new line
End If


If StartCell = "JJC1" Then

Code = "If Jump Switch C is set to 1, jump to location SS"

   If Range("C17").Value = 1 Then ProgramStep = SS          'Jump to new line
End If


If StartCell = "JANZ" Then

Code = "Conditional jump: go to location SS if Register A does not equal 0"
```

```
Set ValueJ = SS

If Range("K2").Value <> 0 Then ProgramStep = ValueJ Else ProgramStep =
ProgramStep + 1

End If


If StartCell = "JBNZ" Then

Code = "Conditional jump: go to location SS if Register B does not equal 0"

Set ValueJ = SS

If Range("K4").Value <> 0 Then ProgramStep = ValueJ Else ProgramStep =
ProgramStep + 1

End If


If StartCell = "JINZ" Then

Code = "Conditional jump: go to location SS if the Index Counter does not
equal 0"

Set ValueJ = SS

If Range("C4").Value <> 0 Then ProgramStep = ValueJ Else ProgramStep =
ProgramStep + 1

End If


If StartCell = "JAZE" Then

Code = "Conditional jump: go to location SS if Register A equals 0"

Set ValueJ = SS

If Range("K2").Value = 0 Then ProgramStep = ValueJ Else ProgramStep =
ProgramStep + 1

End If


If StartCell = "JBZE" Then

Code = "Conditional jump: go to location SS if Register B equals 0"

Set ValueJ = SS

If Range("K4").Value = 0 Then ProgramStep = ValueJ Else ProgramStep =
ProgramStep + 1

End If


If StartCell = "JIZE" Then

Code = "Conditional jump: go to location SS if the Index Counter equals 0"

Set ValueJ = SS
```

```
If Range("C4").Value = 0 Then ProgramStep = ValueJ Else ProgramStep =
ProgramStep + 1

End If
```

The following routine swaps the numbers in Registers A and B (SWAP):

```
If StartCell = "SWAP" Then

Code = "Switch the numbers in Registers A and B"

Range("F120").Value = Range("K2")
Range("F121").Value = Range("K4")

        Range("K2").Value = Range("F121")      'Send value of RegisterA to
        RegisterB

        Range("K4").Value = Range("F120")      'Send value of RegisterB to
        RegisterA

ProgramStep = ProgramStep + 1

End If
```

The following routines replace the number in Register A or B with its square root (SQTA/SQTB):

```
If StartCell = "SQTA" Then

Code = "Take the square root of the number in Register A, placing the result
back into Register A"

Set RegisterA = Range("K2")

        Range("K2").Value = Sqr(RegisterA)     'Send value of RegisterA to
        its square root

   ProgramStep = ProgramStep + 1

End If


If StartCell = "SQTB" Then

Code = "Take the square root of the number in Register B, placing the result
back into Register B"

Set RegisterB = Range("K4")

        Range("K4").Value = Sqr(RegisterB)     'Send value of RegisterA to
        its square root

    ProgramStep = ProgramStep + 1

End If
```

The following routine increments the program step by one unit (NOOP):

```
If StartCell = "NOOP" Then
```

```
Code = "Increase the Program Step by 1, but don't do anything else"
    'No operation
    ProgramStep = ProgramStep + 1
End If
```

The following routine performs an unconditional jump (JUMP):

```
If StartCell = "JUMP" Then
Code = "Unconditional jump to location SS"
    ProgramStep = SS                        'Jump to new line
End If
```

The following routine performs exponentiation (EXPA/EXPB):

```
If StartCell = "EXPA" Then
Code = "Evaluate (Register A)^(value in SS)"
Set RegisterA = Range("K2")
TempSS = Range("F" & (SS + 13)).Value
RegisterA = RegisterA ^ TempSS
        Range("K2").Value = RegisterA
    ProgramStep = ProgramStep + 1
End If


If StartCell = "EXPB" Then
Code = "Evaluate (Register B)^(value in SS)"
Set RegisterB = Range("K4")
TempSS = Range("F" & (SS + 13)).Value
RegisterB = RegisterB ^ TempSS
        Range("K4").Value = RegisterB
    ProgramStep = ProgramStep + 1
End If
```

The following routine sums the numbers in the index counter and the storage location (INDA):

```
If StartCell = "INDA" Then
```

```
Code = "Sum the numbers in the Index Counter and location SS, placing the
result into the Index Counter"

Set IndexC = Range("C4")

IndexC = IndexC + Range("F" & (SS + 13)).Value

        Range("C4").Value = IndexC

    ProgramStep = ProgramStep + 1

End If
```

The following routine replaces the number in the Index Counter with the number in the storage location (INDL):

```
If StartCell = "INDL" Then

Code = "Replace the number in the Index Counter with the number in location
SS"

Set IndexC = Range("C4")

IndexC = Range("F" & (SS + 13)).Value

        Range("C4").Value = IndexC

    ProgramStep = ProgramStep + 1

End If
```

The following routine takes the Index Counter and subtracts the number in the storage location (INDS):

```
If StartCell = "INDS" Then

Code = "Take Index Counter minus the number in location SS, placing the
result into the Index Counter"

Set IndexC = Range("C4")

IndexC = IndexC - Range("F" & (SS + 13)).Value

        Range("C4").Value = IndexC

    ProgramStep = ProgramStep + 1

End If
```

The following routines reads the values from Input A or B and sends that data to the storage location (ENIA/ENIB):

```
If StartCell = "ENIA" Then

Code = "Read the value of Input A, and send that data to location SS"

Set TempInput = Range("G2")
```

```
            Range("F" & (SS + 13)).Value = TempInput    'Send value of input to
                storage

    ProgramStep = ProgramStep + 1

End If


If StartCell = "ENIB" Then

Code = "Read the value of Input B, and send that data to location SS; since B
has multiple inputs possible, increment the 'input tape' as well"

Set TempInput = Range("G" & InputBRow)

            Range("F" & (SS + 13)).Value = TempInput    'Send value of input to
                storage

        InputBRow = InputBRow + 1

        If InputBRow > 9 Then InputBRow = 4

    ProgramStep = ProgramStep + 1
End If
```

The following routines finds the difference between Register A or B and the storage location (SUBA/SUBB):

```
If StartCell = "SUBA" Then

Code = "Take Register A minus the number in location SS, placing the result
into the Register A"

Set RegisterA = Range("K2")

TempSS = Range("F" & (SS + 13)).Value

RegisterA = RegisterA - TempSS

            Range("K2").Value = RegisterA

    ProgramStep = ProgramStep + 1

End If


If StartCell = "SUBB" Then

Code = "Take Register B minus the number in location SS, placing the result
into the Register B"

Set RegisterB = Range("K4")

TempSS = Range("F" & (SS + 13)).Value

RegisterB = RegisterB - TempSS

            Range("K4").Value = RegisterB

    ProgramStep = ProgramStep + 1
```

```
End If
```

The following routines find the products of values in Register A or B and the number in the storage location (MULA/MULB):

```
If StartCell = "MULA" Then

Code = "Take Register A times the number in location SS, placing the result into the Register A"

Set RegisterA = Range("K2")

TempSS = Range("F" & (SS + 13)).Value

RegisterA = RegisterA * TempSS

        Range("K2").Value = RegisterA

    ProgramStep = ProgramStep + 1
End If


If StartCell = "MULB" Then

Code = "Take Register B times the number in location SS, placing the result into the Register B"

Set RegisterB = Range("K4")

TempSS = Range("F" & (SS + 13)).Value

RegisterB = RegisterB * TempSS

        Range("K4").Value = RegisterB

    ProgramStep = ProgramStep + 1
End If
```

The following routines find the quotients of values in Register A or B and the number in the storage location (DIVA/DIVB):

```
If StartCell = "DIVA" Then

Code = "Take Register A divided by the number in location SS, placing the quotient into the Register A and the remainder into Register B"

Set RegisterA = Range("K2")

TempSS = Range("F" & (SS + 13)).Value

Remainder = RegisterA Mod TempSS

RegisterA = RegisterA / TempSS

        Range("K2").Value = Int(RegisterA)

        Range("K4").Value = Remainder
```

```
            ProgramStep = ProgramStep + 1

End If


If StartCell = "DIVB" Then

Code = "Take Register B divided by the number in location SS, placing the
quotient into the Register B and the remainder into Register A"

Set RegisterB = Range("K4")

TempSS = Range("F" & (SS + 13)).Value

Remainder = RegisterB Mod TempSS

RegisterB = RegisterB / TempSS

        Range("K4").Value = Int(RegisterB)

        Range("K2").Value = Remainder

    ProgramStep = ProgramStep + 1

End If
```

The following routines take the decimal number in Register A or B and divide it by the number in the storage location (DVDA/DVDB):

```
If StartCell = "DVDA" Then

Code = "Take the decimal number in Register A and divide it by the decimal
number in location SS, placing the quotient into the Register A"

Set RegisterA = Range("K2")

TempSS = Range("F" & (SS + 13)).Value

RegisterA = RegisterA / TempSS

        Range("K2").Value = RegisterA

    ProgramStep = ProgramStep + 1

End If


If StartCell = "DVDB" Then

Code = "Take the decimal number in Register B and divide it by the decimal
number in location SS, placing the quotient into the Register B"

Set RegisterB = Range("K4")

TempSS = Range("F" & (SS + 13)).Value

RegisterB = RegisterB / TempSS

        Range("K4").Value = RegisterB

    ProgramStep = ProgramStep + 1
```

End If

The following routines display a mixed fraction in a variety of ways:

```
If StartCell = "PABA" Then

Code = "Displays a mixed fraction in Output A; specifically, the whole number
part comes from Register A, the numerator comes from Register B, and the
denominator comes from location SS"

Set RegisterA = Range("K2")

Set RegisterB = Range("K4")

Denom = Range("F" & (SS + 13)).Value

        Range("M" & OutputRowA).Value = RegisterA & " and " & RegisterB &
        "/" & Denom

        OutputRowA = OutputRowA + 1

    ProgramStep = ProgramStep + 1

End If


If StartCell = "PABB" Then

Code = "Displays a mixed fraction in Output B; specifically, the whole number
part comes from Register A, the numerator comes from Register B, and the
denominator comes from location SS"

Set RegisterA = Range("K2")

Set RegisterB = Range("K4")

Denom = Range("F" & (SS + 13)).Value

        Range("O" & OutputRowB).Value = RegisterA & " and " & RegisterB &
        "/" & Denom

        OutputRowB = OutputRowB + 1

    ProgramStep = ProgramStep + 1

End If


If StartCell = "PBAA" Then

Code = "Displays a mixed fraction in Output A; specifically, the whole number
part comes from Register B, the numerator comes from Register A, and the
denominator comes from location SS"

Set RegisterA = Range("K2")

Set RegisterB = Range("K4")

Denom = Range("F" & (SS + 13)).Value

        Range("M" & OutputRowA).Value = RegisterB & " and " & RegisterA &
        "/" & Denom
```

```
        OutputRowA = OutputRowA + 1

    ProgramStep = ProgramStep + 1

End If


If StartCell = "PBAB" Then

Code = "Displays a mixed fraction in Output B; specifically, the whole number
part comes from Register B, the numerator comes from Register A, and the
denominator comes from location SS"

Set RegisterA = Range("K2")

Set RegisterB = Range("K4")

Denom = Range("F" & (SS + 13)).Value

        Range("O" & OutputRowB).Value = RegisterB & " and " & RegisterA &
        "/" & Denom

        OutputRowB = OutputRowB + 1

    ProgramStep = ProgramStep + 1

End If
```

The following routines find the digital roots of Register A or B (DRTA/DRTB):

```
If StartCell = "DRTA" Then

Code = "Find the digital root of the number in Register A, placing the result
into Register A"

Number = Range("K2").Value

Number = Trim(Str(Number))

while (Val(Number) / 10) > 1

    sum = 0

    For i = 1 To Len(Number)

        sum = sum + Val(Mid(Number, i, 1))

    Next

    Number = Trim(Str(sum))

Wend

Range("K2").Value = Number

ProgramStep = ProgramStep + 1

End If
```

```
If StartCell = "DRTB" Then

Code = "Find the digital root of the number in Register B, placing the result
into Register B"

Number = Range("K4").Value

Number = Trim(Str(Number))

While (Val(Number) / 10) > 1

    sum = 0

    For i = 1 To Len(Number)

        sum = sum + Val(Mid(Number, i, 1))

    Next

    Number = Trim(Str(sum))

Wend

Range("K4").Value = Number

ProgramStep = ProgramStep + 1

End If
```

The following routines completely reverse the digits of the numbers in Register A or B (REVA/REVB):

```
If StartCell = "REVA" Then

Code = "Completely reverse the digits of the number in Register A"

s = Range("K2").Value

Number = Trim(s)

NewNumber = 0

Power = 0

Length = Len(Number)

For i = 1 To Length

    x = Val(Mid(Number, i, 1))

    NewNumber = NewNumber + x * (10 ^ Power)

    Power = Power + 1

Next i

Range("K2").Value = NewNumber

ProgramStep = ProgramStep + 1

End If
```

```
If StartCell = "REVB" Then

Code = "Completely reverse the digits of the number in Register B"

s = Range("K4").Value

Number = Trim(s)

NewNumber = 0

Power = 0

Length = Len(Number)

For i = 1 To Length

    x = Val(Mid(Number, i, 1))

    NewNumber = NewNumber + x * (10 ^ Power)

    Power = Power + 1

Next i

Range("K4").Value = NewNumber

ProgramStep = ProgramStep + 1

End If
```

If the debugger option is turned on, once a line of code is run, a dialog box pops up explaining the code.

```
'Debugger, Part 2: Next line is the post-run explanation

If Range("K8").Value = 1 Then MsgBox ("Just ran: LINE " & ProgramStep & ": " & Range("F" & Rowct) & ", " & Range("G" & Rowct) & ". Which means: " & Code & ".")

'Repaint line of code with a clear coat:

 Range("E" & Rowct).Select
   With Selection.Interior
        .Pattern = xlNone
        .TintAndShade = 0
        .PatternTintAndShade = 0
    End With
```

Here's code for the "slow down" option, if it's selected (by typing a 1 into cell L29):

```
'If Pause between-steps is selected, pause:

If Range("k7").Value = 1 Then For Pause = 1 To 100000000: Next Pause
```

Various housekeeping issues wrap up the algorithm:

```
'Increment, getting ready for next program line:
```

```
Rowct = 13 + ProgramStep

Range("C2").Value = ProgramStep

Loop

End Sub
```

Finally, it's time to insert the Compile Program macro. Right click the "COMPILE PROGRAM" button, select "Assign Macro," and click the "New" button. Underneath the subroutine called `Sub Button3_Click()`, type the following:

```
'Declare important variables:

Dim InputValue As Variant
Dim Rowct As Integer
Dim VariableName(1 To 100) As String
Dim VariableValue(1 To 100) As String
Dim VariableSS(1 To 100) As Integer
Dim LabelName(1 To 100) As Variant
Dim LabelLocation(1 To 100) As Integer
Dim Mnemonic As String
Dim IsIttheEnd As Integer
Dim CountWithoutDAT As Integer
Dim ReadWriteSS As Integer

'Initalize important variables

Rowct = 1
IsIttheEnd = 0
CountWithoutDAT = 1
ReadWriteSS = 90


'Clear all inputs/outputs:

    Range("C2").Select
    ActiveCell.FormulaR1C1 = "0"
    Range("C4").Select
    ActiveCell.FormulaR1C1 = "0"
    Range("C9").Select
    ActiveCell.FormulaR1C1 = "="
    Range("C10").Select
    ActiveCell.FormulaR1C1 = "="
    Range("C15").Select
    ActiveCell.FormulaR1C1 = "0"
    Range("C16").Select
    ActiveCell.FormulaR1C1 = "0"
    Range("C17").Select
    ActiveCell.FormulaR1C1 = "0"
    Range("G2:G6").Select
    Selection.ClearContents
    Selection.ClearContents
    Range("K2").Select
    ActiveCell.FormulaR1C1 = "0"
    Range("K4").Select
    ActiveCell.FormulaR1C1 = "0"
    Range("M3:O5000").Select
    Selection.ClearContents
    Selection.ClearContents
    Selection.ClearContents
```

```
    Selection.ClearContents
    Selection.ClearContents
    Range("K25").Select
    ActiveWindow.SmallScroll Down:=93
    Range("F103:F112").Select
    Selection.ClearContents
    Selection.ClearContents
    Selection.ClearContents
    Selection.ClearContents
    Selection.ClearContents
    Range("A1").Select
    Range("G5:G9").Select
    Selection.ClearContents
    Range("A1").Select


    'Clear any highlighted lines of code:

    Range("E13:E112").Select

    With Selection.Interior
        .Pattern = xlNone
        .TintAndShade = 0
        .PatternTintAndShade = 0
    End With

     Range("F13:H112").Select
     Selection.ClearContents

   'Move cursor to top left cell of assembly program:

    Range("Q1").Select


'Count how many lines--excluding lines with DAT at end--there are in the
program:

While Range("R" & CountWithoutDAT).Value <> "DAT"

    CountWithoutDAT = CountWithoutDAT + 1

Wend


'Check to see if any "DAT" statements are used; if so, set the variable name
into the array and assign the initial value (if any) next to the variable
name:

A = 1
For X = 1 To 98
    If Range("R" & X).Value = "DAT" Then
        VariableName(A) = Range("Q" & X)
        VariableValue(A) = Range("S" & X)
        A = A + 1
    End If
Next X



'Output all DAT (data) lines of code to the correct storage locations

For Z = 1 To 98

    If VariableValue(Z) <> "" Then
```

```
        Range("F" & CountWithoutDAT + 12).Value = VariableValue(Z)

        VariableSS(Z) = CountWithoutDAT - 1

        CountWithoutDAT = CountWithoutDAT + 1

    End If

    If VariableValue(Z) = "" And VariableName(Z) <> "" Then

        VariableSS(Z) = ReadWriteSS

        ReadWriteSS = ReadWriteSS + 1

            If ReadWriteSS > 99 Then ReadWriteSS = 90        'There are only
                10 read-write storage locations available; if the number of
                read-write variables exceeds that, they need to be reassigned
                overtop of older variables--which will certainly cause an error
                in the execution of the program. So don't create more than ten
                read-write variables!

    End If

Next Z


'Check to see if there are any labels:

For X = 1 To 98
    If Range("Q" & X).Value <> "DAT" And Range("Q" & X).Value <> "" Then
        LabelName(X) = Range("Q" & X)
        LabelLocation(X) = X
    End If
Next X


'The loop to translate the program into machine language starts here:

While IsIttheEnd = 0        'Until the variable "IsIttheEnd" equals 1, the
loop continues....


'Get mnemonic

Mnemonic = Range("R" & Rowct)


'Output the line of code to the correct storage location

If Mnemonic <> "DAT" Then

    Range("F" & Rowct + 12).Value = Mnemonic

    'Search for operand

    If Range("S" & Rowct).Value = "IN" Then Range("G" & Rowct + 12).Value =
    100

    If Range("S" & Rowct).Value <> "IN" Then    'Search first for label; if
    found, output label storage location

        For Y = 1 To 98
```

```
            If Range("S" & Rowct).Value = LabelName(Y) Then Range("G" & Rowct
            + 12).Value = LabelLocation(Y) - 1

        Next Y

    'Output the correct variable storage locations:

    For Z = 1 To A

        If Range("S" & Rowct).Value = VariableName(Z) Then Range("G" & Rowct
            + 12).Value = VariableSS(Z)

    Next Z


    'Fix several mnemonics to have no operand:

    If Mnemonic = "STOP" Or Mnemonic = "NOOP" Then Range("G" & Rowct +
    12).Value = ""

        End If

End If


Rowct = Rowct + 1

'End the program upon encountering the first blank line

If Range("Q" & Rowct).Value = "" And Range("R" & Rowct).Value = "" And
Range("S" & Rowct).Value = "" Then IsIttheEnd = 1

Wend


'Move cursor to top left cell of worksheet:

    Range("A1").Select

End Sub
```

a simple example program

Our machine language program example below is designed to run on the Instructo emulator. Before typing in the program, click the "RESET/CLEAR" button (you may want to make a habit of pressing this button before every program run), and then enter two numbers into Input B (in cells G4 and G5). Notice that there are, at a maximum, six cells available for input— seven, if you include the numerical-data-only Input A.

Instructo Program No. 8-1: Adding and Subtracting

LINE	MNEMONIC	SS/IN	COMMENTS
00	ENIB	90	Enter information from Input B into storage location 90.
01	ENIB	91	Enter information from Input B into storage location 91.

02	ADDA	90	Add to the number in Register A the number in storage location 90.
03	ADDA	91	Add to the number in Register A the number in storage location 91.
04	SUBA	08	Subtract from the number in Register A the number in storage location 08.
05	STRA	92	Store the contents of Register A into storage location 92.
06	PROA	92	Print in Output A the contents of storage location 92.
07	STOP	07	Stop the program, leaving the program step indicator on line 07.
08	10		Place the number 10 into read-only storage.

Once the code is typed in, click the "START/STOP" button to run the program. The program adds the two inputs together, subtracts 10, and then outputs the result in Output A.

Keep in mind that the Instructo only designates storage locations 90 to 99 read/write status (i.e., "Main Storage"); all the remaining storage locations are read only. In addition, recall that the Program Step Counter and Index Counter are both *always* set to zero before running a program.

Also realize that the Instructo syntax in this guide— and emulator— is not identical to that shown in the Instructo's operator's manual. For instance, an unconditional jump to storage location 57 would appear as JUMP, 57 in the operator's manual, but you would enter in JUMP 57 into the emulator, dispensing with the comma.

There's yet another caveat, this time dealing with the syntax of the Excel emulator. If IN, a reference to the Index Counter, is the operand, you *must* instead enter in 100 as the operand.

Finally, if necessary, you can interrupt the macro at any time by repeatedly pressing the Escape key.

Working with the complier, which will translate assembly language into machine code, is straightforward, as long as you follow the instructions above the "COMPILE PROGRAM" button. Be especially careful not to initialize more than ten read-write variables in your program.

Type in the following assembly program (be careful not to mis-key anything, since the compiler offers no error-checking mechanism).[*]

Instructo Program No. 8-2: Adding and Subtracting Redux

LABEL	MNEMONIC	OPERAND	COMMENTS
start	ENIB	first	Stores in the first user input as the variable "first."
	ENIB	second	Stores in the second user input as the variable "second."

---

[*] Unlike the assembler for the CARDIAC/LMC, which has a simple error check: with mis-keyed instructions, the program will be on the fritz.

	ADDA	first	Adds the "first" variable to the value in Register A.
	ADDA	second	Adds the "second" variable to the value in Register A.
	SUBA	ten	Subtracts the value of the "ten" variable (which is 10) by the value in Register A.
	STRA	result	Stores the value in Register A into a variable called "result."
	PROA	result	Outputs the value of "result."
	STOP		Ends the program.
ten	DAT	10	Initialize a read-only variable.
first	DAT		Initialize a read-write variable.
second	DAT		Initialize a read-write variable.
result	DAT		Initialize a read-write variable.

After pressing the "COMPILE PROGRAM" button, you'll see a machine language version of the program copied into the correct storage locations, all ready to run. As expected, the machine code produced is effectively identical to Instructo Program 8-1.

Occasionally, after you run a compiled assembly program once or twice, the next run locks the machine up. To avoid this complication, click the "COMPILE PROGRAM" button prior to *every* run of the assembly program.

section

a turing machines emulator

A Turing machine, first formulated by mathematician Alan Turing, is a finite-state machine with a read/write head capable of processing an instruction set on a blisteringly long paper tape. The tape is divided into squares with printed symbols, such as 0 or 1 (or nothing at all).

Running an instruction set on a Turing machine is often tedious without a computer program to assist you. Below you will find such a program, in the form of an Excel macro. Although there are a number of such virtual Turing machines online, ours will be especially easy to program (no scripting language required)— once we get it up and running.

getting started
First, you'll need to make sure that you have the "Developer" tab in the Excel "Ribbon" (the strip of options right below the menu) available. Go to "File," then "Options," then "Customize Ribbon." Finally, check off the "Developer" option. Even if you don't have a "Ribbon," you'll need to make sure that the "Developer" option is selected in order to proceed.

Open up a new Excel spreadsheet, and then save it with the filename "TURING." Make sure your workbook is "Macro-Enabled" when saving.

setting up the worksheet
First, right click the worksheet's tab name, type "TURING," and then click back inside the worksheet.

Highlight cells A3 to P4, and change their size to 20-point font. Next, highlight cells B3 to P3 and insert borders around each individual cell. Type one "B" into cell, from cells B3 to P3, and then type a caret (^) symbol into cell F4.

In cell R1, type "FULL TAPE." Underline and bold cell R1 as well. In cell A6, type "Steps performed so far:" and in cell C6, type the number 0.

Cells A9 to A17 will contain a list of instructions. Match the screenshot below on your Excel worksheet:

6	Steps performed so far:	0
7		
8		
9	ENTER INSTRUCTION SET BELOW	
10	Type B for a blank space on the paper tape.	
11	Type N for no movement left or right.	
12	Type L to move left, R to move right.	
13	Each State needs three rows: a row for B, for 0, and for 1, in that order.	
14	Type 0 to represent Halt.	
15	If not using a State, leave it blank.	
16	Before running program, enter in input on the TAPE line above (the read/write	
17	head is represented by the caret).	
18		

Underneath the instructions will lie the instruction set— i.e., the space to enter in the Turing machine program. If you wish, you can make space for hundreds of states; however, for our purposes here, we'll only set up a three-state machine. Starting in cell A21, match the following screenshot on your worksheet:

	STATE	READ	WRITE	MOVE	GO TO STATE
21	STATE	READ	WRITE	MOVE	GO TO STATE
22	1	B			
23	1	0			
24	1	1			
25	2	B			
26	2	0			
27	2	1			
28	3	B			
29	3	0			
30	3	1			
31					

Before continuing, make sure your worksheet resembles the following:

coding the macro

We will first need to insert a button that will run our macro. Click on the "Developer" tab, select the "Insert" dropdown, and click the gray button. Draw a rectangular button, starting around cell I18; once you let go of the mouse button, a dialog box will pop up asking you which macro you'd like to assign to the button. Click "Cancel" for now, and instead right click the button and select "Edit Text." Type "RUN INSTRUCTION SET" and click somewhere outside of the button. Your worksheet should now resemble this screenshot:

It's now time to insert the macro. Right click the "START" button, select "Assign Macro," and click the "New" button. Underneath the subroutine called `Sub Button1_Click()`, type the following:

```
'Run Program macro

Dim Tape(200) As String
Dim CurrentSymbol As String
Dim NextState As Integer
Dim CurrentMove As String
```

The `Dim` statements above declare some of the important variables for later use. Some variables are of type `String`, meaning that characters of any sort can be stored in them, while others are of type `Integer`, which is self-explanatory. There is also one array initialized, called `Tape`, which keeps track of the symbols on the paper tape.

Sprinkled throughout the code are comments— shown as phrases that begin with apostrophes— that have been added for clarity.

Next, some variables are set to initial values:

```
Halt = 0       'The Halt variable is set to zero

BeginningofTape = 0

Rowct = 22         'Current row
```

```
ReadSymbol = 0    'Current read symbol (0 means blank, 1 means zero, and 2
means one)

NumberofSteps = 0

Range("C6").Value = NumberofSteps    'Resets number of steps
```

The input values on the paper tape need to be read into the Tape array:

```
'Initialize the first 200 elements of the Tape array to be blank:

For X = 1 To 200

    Tape(X) = "B"

Next X


'Capture initial input from tape into the Tape array:

For Y = 66 To 80

    Tape(Y) = Range(Chr$(Y) & 3).Value

Next Y
```

Next begins a `While...Wend` loop to decode and run the Turing machine program. Inside the loop lies the code to interpret the instruction set, as well as message boxes that pop up to alert the user to each step in the process.

```
While Halt <> 1

'Display symbols on paper tape

For D = 66 To 80

    Range(Chr$(D) & 3).Value = Tape(D + BeginningofTape)

Next D

For C = 2 To 201

    Range("R" & C).Value = Tape(C - 1)

Next C

CurrentSymbol = Range("F3")

'Set CurrentSymbol to correct ReadSymbol

If CurrentSymbol = "B" Then ReadSymbol = 0
If CurrentSymbol = "0" Then ReadSymbol = 1
If CurrentSymbol = "1" Then ReadSymbol = 2


If CurrentSymbol = Range("B" & (Rowct + ReadSymbol)) Then

    'Write the new symbol to the cell, and save it in the Tape array; then
    re-display tape:
```

```
        WriteSymbol = Range("C" & (Rowct + ReadSymbol)).Value

        Range("A" & Rowct + ReadSymbol & ":E" & Rowct + ReadSymbol).Select

        MsgBox ("The current symbol is: " & CurrentSymbol & ". Ready to WRITE
        the symbol " & WriteSymbol & ".")

        Range("F3").Value = WriteSymbol

        Tape(66 + BeginningofTape + 4) = WriteSymbol

        For D = 66 To 80

            Range(Chr$(D) & 3).Value = Tape(D + BeginningofTape)

        Next D

        For C = 2 To 201

            Range("R" & C).Value = Tape(C - 1)

        Next C

      'Move the tape either left or right (or not at all)

        CurrentMove = Range("D" & (Rowct + ReadSymbol)).Value

        If CurrentMove = "L" Then

            BeginningofTape = BeginningofTape - 1

        End If

        If CurrentMove = "R" Then

            BeginningofTape = BeginningofTape + 1

        End If

      'Determine which state to go to next--or if it's time to halt:

        NextState = Range("E" & (Rowct + ReadSymbol))

            If NextState = 0 Then Halt = 1 Else Rowct = 19 + NextState * 3

        'Add one to the number of steps counter.

        NumberofSteps = NumberofSteps + 1

        Range("C6").Value = NumberofSteps

        MsgBox ("Ready to READ next symbol; but first, will travel " &
        CurrentMove & " and then to the following STATE: " & NextState & ".")

    End If

    Wend

    MsgBox ("Machine has HALTED")

End Sub
```

an example program

Let's run a Turing machine program presented earlier in the guide: the 2-State Busy Beaver. Recall the program:

Turing Machine Program No. 9-1: The 2-State Busy Beaver

STATE	READ	WRITE	NEXT MOVE	NEXT STATE
1	Blank	1	R	2
1	0	Blank	None	1
1	1	1	L	2
2	Blank	1	L	1
2	0	Blank	None	1
2	1	1	R	Halt

In order to enter in the program properly, however, you'll need to change the "Halt" to 0, the "Blank" to "B," and the "None" to "N." See the screenshot below.

	STATE	READ	WRITE	MOVE	GO TO STATE
21					
22	1	B	1	R	2
23	1	0	B	N	1
24	1	1	1	L	2
25	2	B	1	L	1
26	2	0	B	N	1
27	2	1	1	R	0
28	3	B			
29	3	0			
30	3	1			

Once the program is keyed in, click the "RUN INSTRUCTION SET" button to run. The emulator will pause with each read-write step, prompting you for a keystroke, and it won't end until four 1s are printed on the paper tape; the emulator will also print the *entirety* of the paper tape off to the side. However, if necessary, you can interrupt the macro at any time by repeatedly pressing the Escape key.

epilogue: completely unfolded

In his book *The Caped Crusade: Batman and the Rise of Nerd Culture*, author Glen Weldon explains how, by the 1970s, "a new breed of enthusiast began to rise to prominence," one that had been quietly lurking in the shadows:

> They called themselves fans, experts, *otaku*. Everyone else, of course, called them nerds....
>
> Nerds had spent decades creating and policing carefully wrought self-identities around their strictly specialized interests [such as comics and computers].... What united them, however, were not the specific objects of their enthusiasm but the nature of the enthusiasm itself— the all-consuming degree to which they rejected the reflexive irony their peers prized.

But what triggers such a deep-seated enthusiasm in adulthood? Weldon doesn't attribute it all to nostalgia, yet he notes that "nostalgia is the nutrient agar upon which all of nerd culture grows."

Indeed, nostalgic reasons, and a need to demythologize my past, led me to start thinking and then researching paper computers nearly a decade ago. I purchased a pristine CARDIAC on eBay and built a replica Instructo— while my passions slowly, inexorably lost any "reflexive irony" in favor of an all-consuming serious. There's no gainsaying: I became a paper computer nerd.

Early on I decided to document my findings, so I wrote a short chapter on paper computers for my first book, *Affront to Meritocracy*. In an all-roads-lead-to-Rome sort of way, seeing connections between everything everywhere, I believed the topic furthered the book's central theme, since, as I put it then, paper computers that relied on human operators in place of electrical components "were... an affront to Moore's Law and the meritocracy of electronic computing." (After all, recall the Instructo operator's manual touting the Instructo as being able to "do almost anything a real electronic computer can do, but it works using pencil and paper instead of electricity and integrated electronic circuits.") In retrospect, however, my zealous pronouncement overstated the matter by betraying a misunderstanding of the key purpose of paper computers: as instructional tools, never as alternatives to microcomputers or, later, desktops. True, paper computers like David Hagelbarger's CARDIAC and Fred Matt's Instructo could run actual programs, albeit with quite a bit of (literal) prodding, but that was not their *raison d'etre*. I once thought it a distinction without a difference, but that's simply not the case: paper computers clearly were never intended as substitutes for electronic ones. To those who were misled by my earlier words, I sincerely apologize. This book has been, among many other things, an attempt to correct the record.

Most important, though, this book has been an answer to a question. My first encounter with the Instructo was in 1989, when the Age of the *Personal Computer* was already in full bloom. My second encounter was over twenty years later, when I was fishing tentatively around the internet for information on it and other paper computers only to stumble upon this message-in-a-bottle post:

> *I invented, patented and published The Instructo Paper Computer when I was a teacher during the early 80's. It was published by McGraw Hill, and I used to be able to find it on the web at some computer museum. It was a pseudo-assembly language device that I used with my math classes. Anybody have any info on it?*

Yes, Mr. Matt. I do.

selected bibliography

Alcosser, E., Phillips, J., & Wolk, A. (1967). *How to build a working digital computer*. New York: Hayden Book Company.

Appleman, D. (1994). *How computer programming works*. Emeryville, Calif.: Ziff-David Press.

Berkeley, E. (1949). *Giant brains, or machines that think*. Hoboken, N.J.: John Wiley & Sons.

Brumbaugh, L., & Yurcik, W. (2005). "The 'little man storage' model." *Proceedings of the 2005 Workshop on Computer Architecture Education Held in Conjunction with the 32nd International Symposium on Computer Architecture— WCAE '05*.

Cone, E. (2000). "Computers, real cheap!" Wired, Apr. 1, 2000. Retrieved from https://www.wired.com/2000/04/paper/

Crawford, M. (2015) *The world beyond your head: On becoming an individual in an age of distraction*. New York: Farrar, Straus and Giroux.

Dijkstra, E. (1988). "On the cruelty of really teaching computing science." *E. W. Dijkstra Archive*. Center for American History, University of Texas at Austin.

Edmundson, M. (2014). *Why teach?: In defense of a real education*. New York: Bloomsbury.

Englander, I. (2014). *The architecture of computer hardware, systems software, and networking: An information technology approach* (5th ed.). Hoboken, N.J.: John Wiley & Sons.

Gertner, J. (2012). *The idea factory: Bell Labs and the great age of American innovation*. New York: Penguin.

Hagelbarger, D. (1956). "SEER, a sequence extrapolating robot." *IRE Transactions on Electronic Computers*, Mar. 1956, 1-7.

Hagelbarger, D., & Fingerman, S. (1968). *An instructional manual for CARDIAC: A cardboard illustrative aid to computation*. Murray Hill, N.J.: Bell Telephone Laboratories.

Hyde, R. (2006). *Write great code, vol. 2: Thinking low-level, writing high-level*. San Francisco: No Starch Press.

Matt, F. (1979). *Instructo paper computer operator's manual: A real programmable paper computer*. New York: McGraw-Hill.

_____ & Moyer, E. (1980). *Programs for Instructo paper computer: Practical application.* New York: McGraw-Hill.

_____ & Moyer, E. (1980). *Programs for Instructo paper computer: Sports calculations.* New York: McGraw-Hill.

_____ & Moyer, E. (1980). *Programs for Instructo paper computer: Geometry.* New York: McGraw-Hill.

Neumann, J. (1993). "First draft of a report on the EDVAC." *IEEE Annals of the History of Computing,* 15(4), 27-75.

Petzold, C. (1999). *Code: The hidden language of computer hardware and software.* Redmond, Washington: Microsoft.

Pitre, B., & Loguidice, B. (2014). *CoCo: The colorful history of Tandy's underdog computer.* Boca Raton, F.L.: CRC Press.

Scott, T. (1962). *Basic computer programming.* New York: Doubleday.

Shannon, C. (1953). "A mind-reading (?) machine." *Bell Laboratories Memorandum,* Mar. 18, 1953.

Turing, A. (1960). "On computable Numbers, with an application to the *Entscheidungsproblem.*" *Annual Review in Automatic Programming,* 230-264.

Stuart, B. (2010). "CARDIAC." Retrieved from https://www.cs.drexel.edu/~bls96/museum/cardiac.html

Weldon, G. (2016). *The caped crusade: Batman and the rise of nerd culture.* New York: Simon & Schuster.

White, R. (1993). *How computers work.* Emeryville, Calif.: Ziff-David Press.

Yurcik, W., & Osborne, H. (2001). "A crowd of Little Man Computers: Visual computer simulator teaching tools." *Proceeding of the 2001 Winter Simulation Conference* (Cat. No.01CH37304).

about the author

MARK JONES LORENZO, who has written books on computer programming, statistics, and the mathematics of lotteries, is a teacher. He lives in Pennsylvania with his dogs.

Made in the USA
Las Vegas, NV
20 September 2024

95555974R00142